全球顶级水疗设计

THE WORLD SPA DESIGN

养生中心 美容中心 酒店水疗 室内设计 1/2

HOTEL SPAS & BEAUTY SPAS \ WELLNESS CENTERS INTERIOR DESIGN

Vol.1

大连理工大学出版社

深圳市艺力文化发展有限公司 编

图书在版编目(CIP)数据

全球顶级水疗设计. 1、2 : 英、汉 / 深圳市艺力文化发展有限公司编. — 大连 : 大连理工大学出版社, 2011.8

ISBN 978-7-5611-6325-2

Ⅰ. ①全… Ⅱ. ①深… Ⅲ. ①保健建筑—建筑设计—作品集—世界 Ⅳ. ①TU246.9

中国版本图书馆CIP数据核字（2011）第134380号

出版发行：大连理工大学出版社
（地址：大连市软件园路80号　邮编：116023）
印　　刷：利丰雅高印刷（深圳）有限公司
幅面尺寸：245mm × 290mm
印　　张：32
插　　页：8
出版时间：2011年8月第 1 版
印刷时间：2011年8月第 1 次印刷
责任编辑：初　蕾
特约编辑：李　莹
责任校对：仲　仁
装帧设计：杨思雁

ISBN 978-7-5611-6325-2
定　　价：560.00元（共两册）

电 话：0411-84708842
传 真：0411-84701466
邮 购：0411-84703636
E-mail：designbook@yahoo.cn
URL：http://www.dutp.cn

如有质量问题请联系出版中心：（0411）84709246　84709043

PREFACE

1

In order to create a memorable spa experience one must address as many of the senses as possible. An empirical experience—all that one can potentially see, hear, touch, taste and smell—drives our spa design and has become the prevailing aspect linking all of our creative work. Spas are a highly specialized building type and require a highly sophisticated set of guest parameters and expectations in comparison to the design of other hospitality spaces. As the nature of the spa continuously evolves, they also serve and provide the proactive remedy to our current cultural needs; primarily stress reduction. The earlier forms as European weight loss clinics have advanced to shape the various needs of people worldwide including destination spa getaways, hotel spas, day spas, wellness centers, exotic resorts, classic thalasso and thermal spas and medical spas. Their significant contribution to the 'sensuous experience' is persistently delightful. A heightened design sensitivity—including acoustics and lighting—is key to the success of these delicate environments. Integrating all five senses in the design of spas guarantees a 'memorable guest experience', reinforcing the connection between mind and body, which is unique to the spa and wellness industry. Let us continue to celebrate the senses through meaningful spa design.

Robert D. Henry
Robert D. Henry Architects

2

The warm and bright spring always comes after the cold long winter. New sprouts and leaves are coming out and everything what people catch into their eyes may start shining again. Then, feeling the pleasure of new lives, the season may be gently moving to the energetic summer. Perhaps SPA should be a place where provides such a sense of wonderful experience. Visiting spas, people may wash out not only their physical fatigue but also mental tiredness. They may find themselves well treated and refreshed from the bottom of hearts in their spa journeys.

Today the wellness business has been expanded the markets to all over the world and so many spas can be recognized in millions of different regions. There are varieties of spa styles, and 'relax, luxury, noble, friendly, exciting, dramatic, natural, ...' lots of words stand to express our spa experiences. However, it is sometimes tough to define what a good spa space is with only a specific word because the style demands can be influenced by and changed in situations, such as regional aspects, cultural background, and economical facts. That is why spa owners and designers have to be very careful whenever they consider a new concept and design for their spa.

There might be some solid and strong tips to understand how to organize and design a successful spa space. It does not matter if a concept is about water or fire, and about European or Asian. Operation and space design have to be linked well to each other and always the spa experiences should be very exciting and dramatic for all coming guests. Spa owner and designers have to imagine how their guests may feel as they come into the spa and stay for treatments. It might be even more important to consider of the future guests, such as 1 year, 5 years and 10 years later. As long as spa is a place to treat people, imagining and caring of guests should be the first step to provide a good space. Then, to build up their own design tips, they better not stay but move to check out what is going on around the world, what problem other owners and designers face to and how they solve it. The only fact about designing spas is 'A good spa has its own successful theory'. As they remember and keep improving, they will someday meet their own best spa.

Takuji Kamio
Kamio Design

10
YMCA Renaissance Center
228
Conti Day Spa
150
Room 20
054
Elysyle Spa
146
Banyan Tree Spa Shanghai
136
Hilton Mexico City Reforma Hotel Spa & Fitness Center

DIRECTORY

186
Spa Al Wadi

Mandi Bali Kobe
Hairu Hair Treatment
162
Art Pelé
The Setai Club at the Setai New York
232
In Yoga Center Spa

104
Home Spa
204
Yoga Deva
Spa Chairama
008
I-WAY
016
SPA Hotel Bründl
242
220
Mira Spa
060
Scandinave Les Bains Vieux
082
Oasis Club
108
Manzi Natural Spa
192
EF MediSpa
034
Aqua Sheko Kensington
130
Nagomi
Vdara Health and Beauty
170
Sasha
114
Kanebo Sensai Select Spa
126
Mandi Bali Nagoya
120
Mars The Salon Spa Boutique

I-WAY

Title: I-WAY
Design Agency: Cyrille Druart
Location: Lyon, France
Client: I-Vent
Date of completion: June 2008
Area: 6500 m^2
Photography: Cyrille Druart

introduction:

Inaugurated in July 2008 in Lyon, France, I-WAY is an international first. This exclusive building is entirely dedicated to automobile simulations (simulator technology offering 6 degrees of freedom inspired by the aeronautics industry), providing the public with access for the first time to eighteen machines. Organised around 3 simulation zones (Formula 1, Endurance, Rally/Touring), the complex also includes numerous upscale areas for a global experience: fitness room, alcohol-free bar, bar/restaurant lounge with terraces, two conference rooms, meeting rooms and offices.

The entrance to the sports room introduces a radical change of environment. While the world of automobiles is shrouded in a dark atmosphere, associated with hard and mechanical materials, the soft whiteness of the reception area represents the human body, its refinement and elegance. The heavy simulators disappear to make way for the lightness of the body. Biology is the opposite of mechanics; the spirit rises and the body floats. Conceptually, this room is a space that defies the laws of gravity.

The path down to the cardio-weight training and fitness areas reveals concrete once again; strength and beauty, effort and sweat. We designed the spa to be as empty and smooth as possible, so that the mind is not burdened by any details and is free from any physical attachments. The space is made up of a series of boxes, internal envelopes inside which people can empty their minds and relax. In this austere place, like a monastery, void of all artifice, the powerful walls reflect our own image, placing us at the center stage.

FORMULE 1
ENDURANCE
RALLYE
EXPERIENCE
(1 COURSE DANS CHAQUE DISCIPLINE)
60€
200€

1

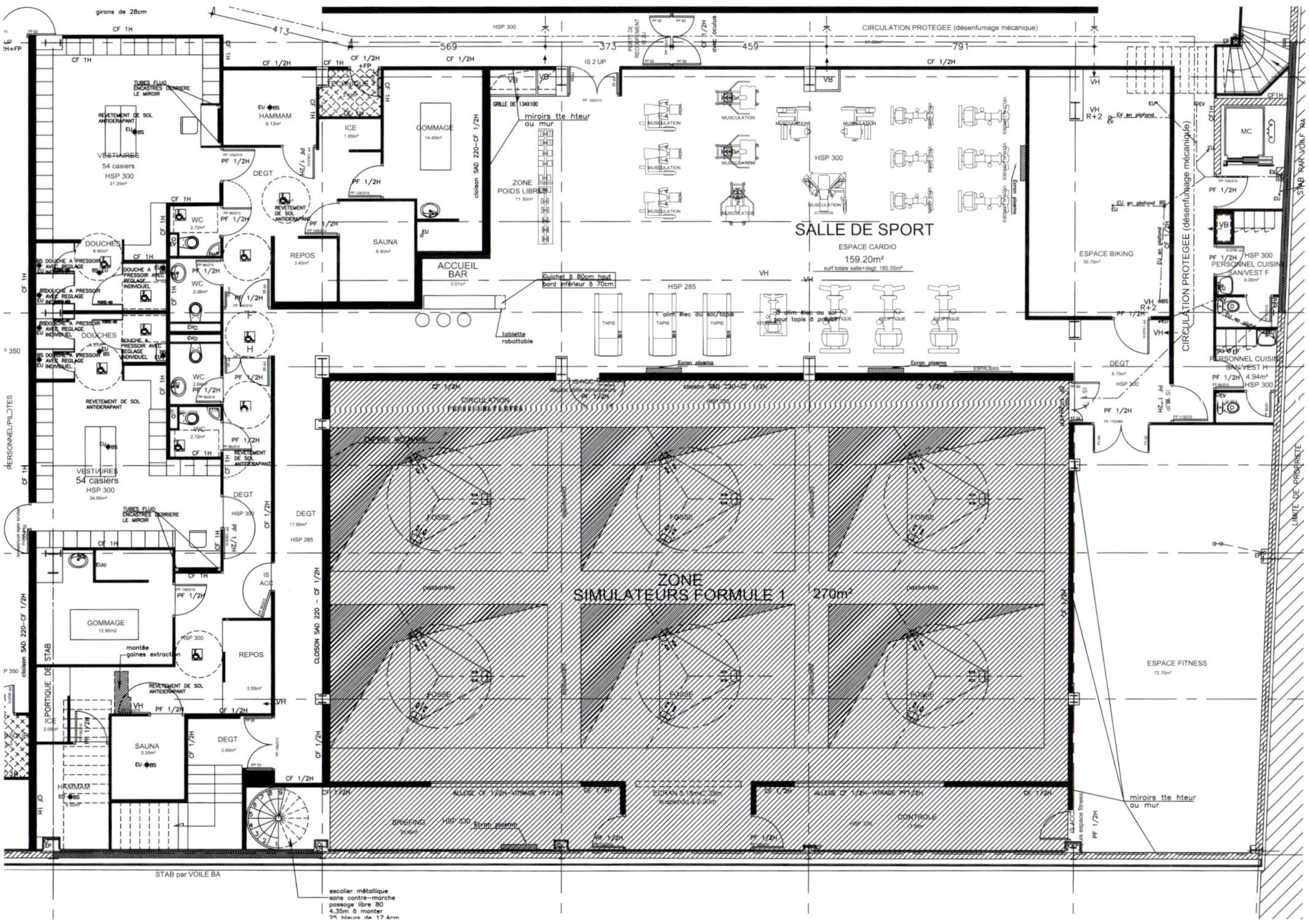

girons de 28cm
CIRCULATION PROTEGEE (désenfumage mécanique)
HAMMAM
ICE
GOMMAGE
VESTIAIRES
54 casiers
HSP 300
DOUCHES
WC
DEGT
REPOS
SAUNA
ACCUEIL
BAR
ZONE
POIDS LIBRES
miroirs tte hteur
ou mur
SALLE DE SPORT
ESPACE CARDIO
159.20m²
ESPACE BIKING
HSP 285
ZONE
SIMULATEURS FORMULE 1
270m²
ESPACE FITNESS
CIRCULATION PROTEGEE (désenfumage mécanique)
PERSONNEL CUISINE
MC
GOMMAGE
montée
gaines extraction
STAB par VOILE BA
escalier métallique
sans contre-marche
passage libre 80

2

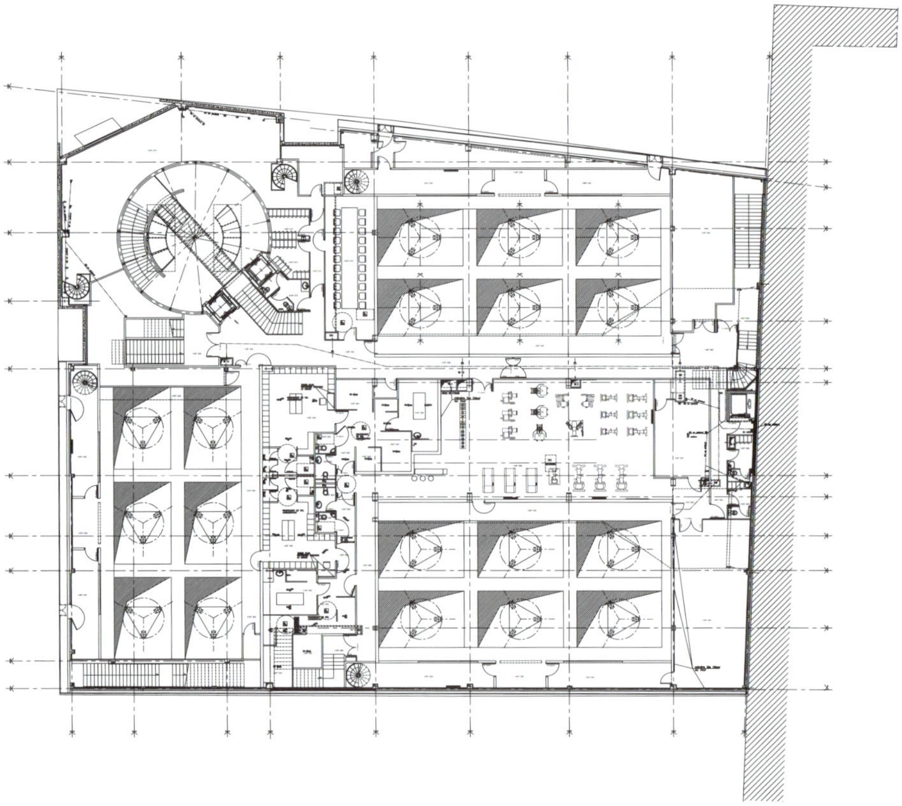

SAUNA
SOINS

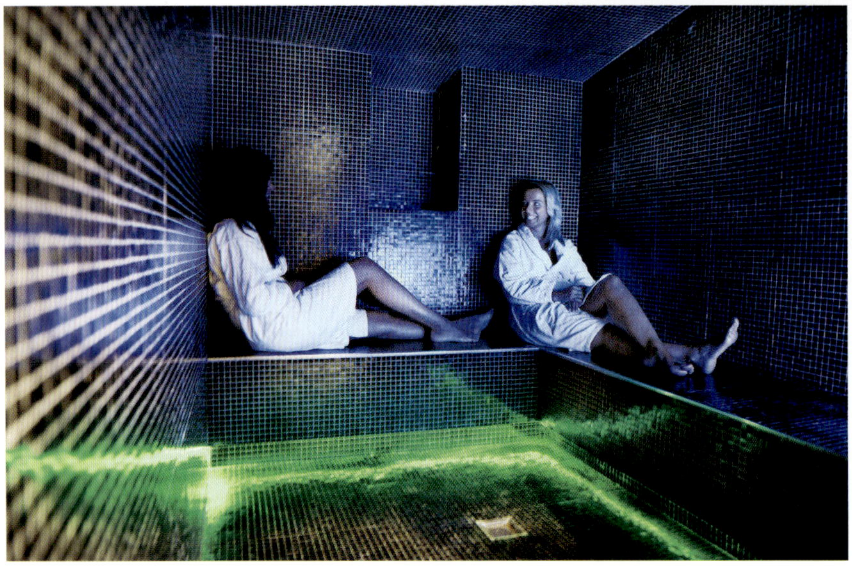

SPA Hotel Bründl

Title: SPA Hotel Bründl
Design Agency: Team M Architects / ISA STEIN STUDIO
Location: Bad Leonfelden
Client: Hypo Impuls Immobilien Leasing GmbH
Area:7898 m^2
Photography: ISA STEIN

introduction:

The building is designed in a contour line and positioned on a sloping hill. It newly defines the piece of land on which it is built. A design which attempts to absorb the pre-defined parameters of the environment and which would appear to 'reinforce' the natural surroundings.

To concentrate 'energy', the building itself has been conceived as a three-dimensional contour line. The building is not erected as a straight structure but rather folded in on itself and twisted out of the axis of its own three-dimensional nature (development of height). A three-storey hall as a connecting link between the two structures which, however, simultaneously has the quality of serving as a multifunctional room and a tool which opens up into the restaurant, bar, therapy zone, rooms and spa.

1

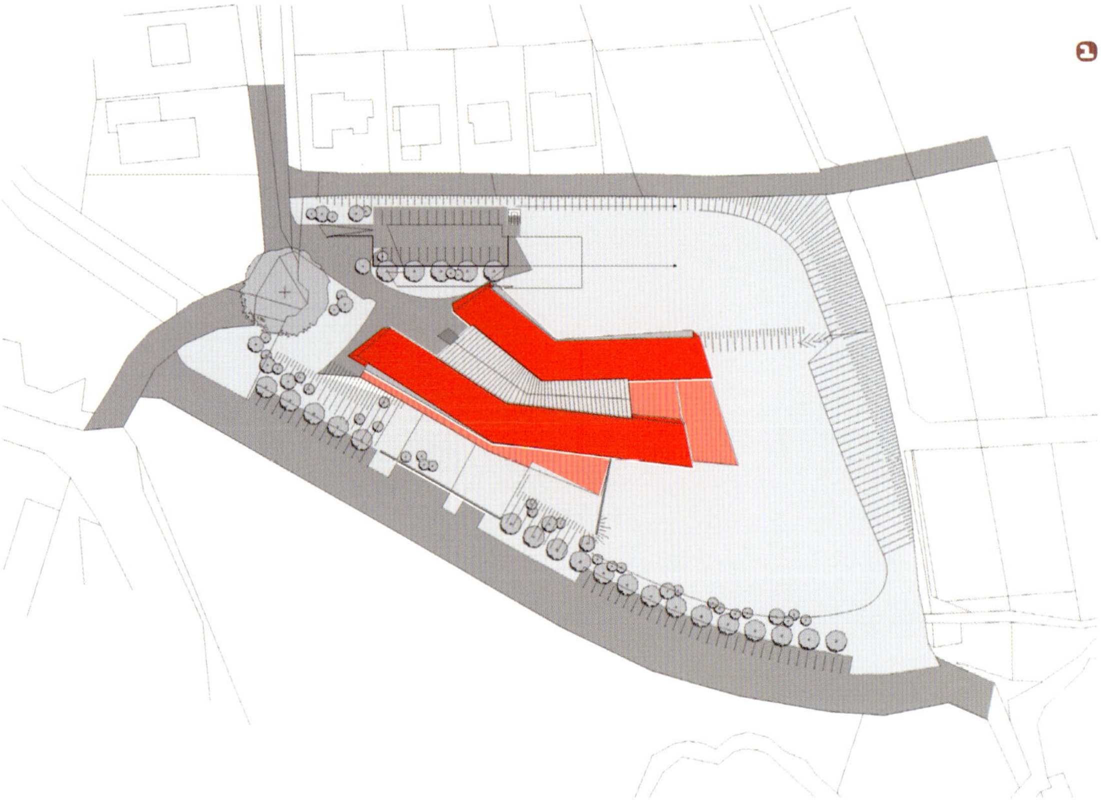

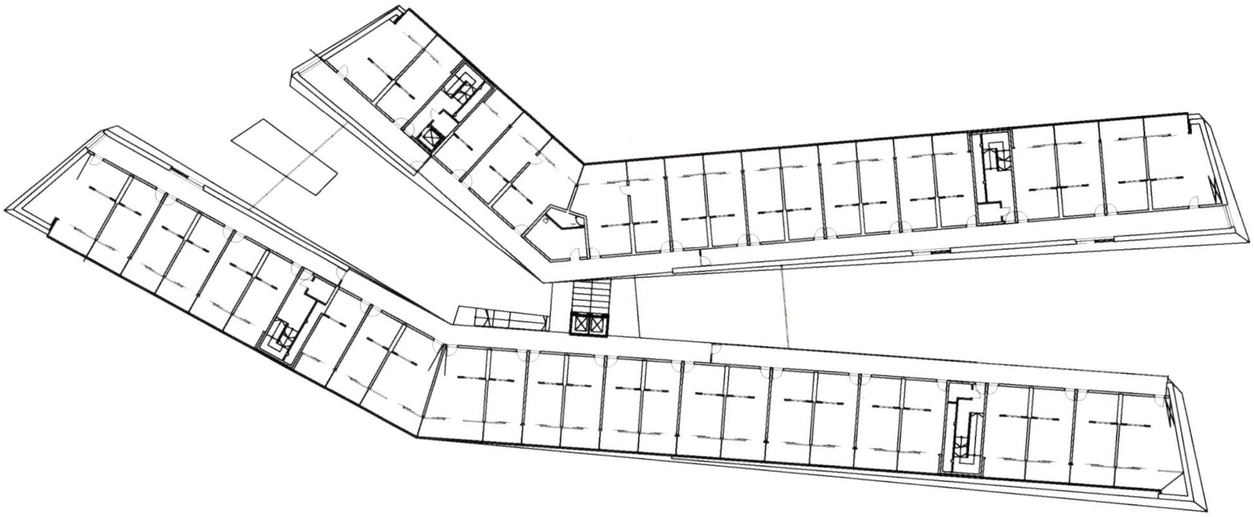

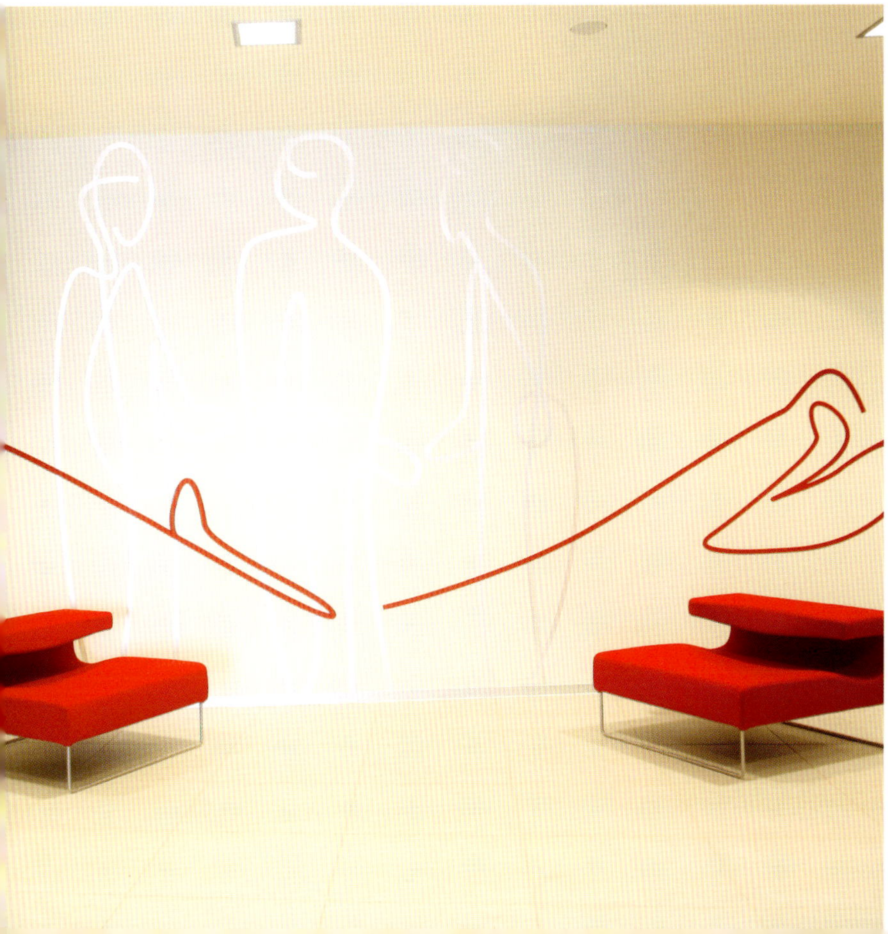

RUHEpol

Spa Chairama

Title: Spa Chairama
Design Agency: Mazzanti Arquitectos; Planb Arquitectos
Location: Bogotá, Colombia
Client: Chairama S.A.
Area:2125 m^2
Photography: Sergio Gómez

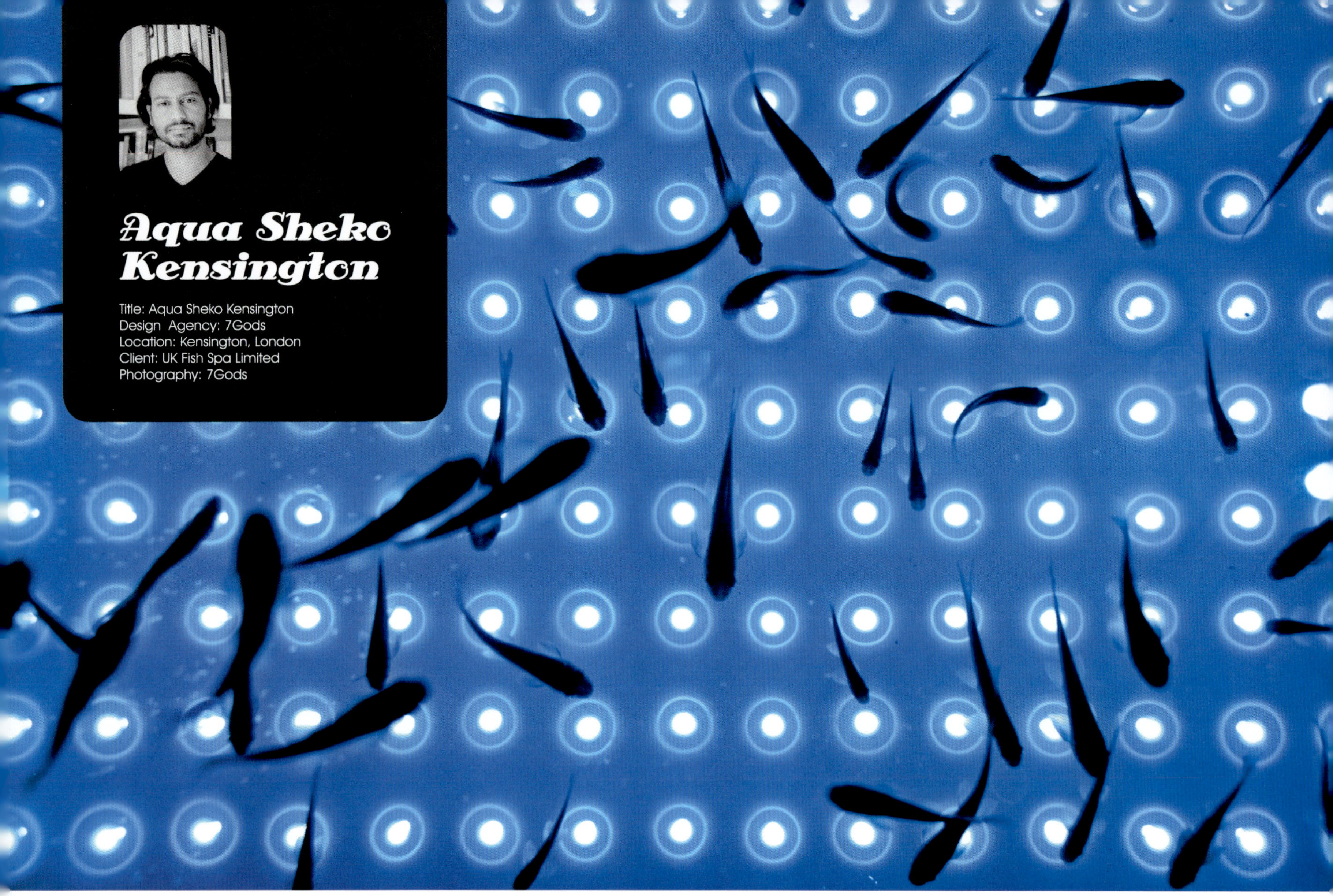

Aqua Sheko Kensington

Title: Aqua Sheko Kensington
Design Agency: 7Gods
Location: Kensington, London
Client: UK Fish Spa Limited
Photography: 7Gods

introduction:

The Aqua Sheko is the unique spa concept designed by up-and-coming London based design agency 7Gods. The aim of the Aqua Sheko is to foster the participation in the unusual treatment experience. The spa introduces the Garra Rufa fish, rare in Europe, this Asian idea promotes a natural way for skin treatment and refreshment.

The first branch was opened in Kensington, London. The design concept as an ambient zone is out of the high street context and conveys escape from the city. The elegant interior uses a dark décor and a monochrome of the dim tones to calm the senses and to relax. The space illumination has been carefully chosen to create a understated and relaxing space. Both lighting and wall pattern are bespoke design by 7Gods. The Aqua Sheko has enabled 7Gods to bring a entire new design experience to London's spa interior design. Kensington spa is highly functional, nonetheless exceedingly appealing to the eye, it shows a great example of where the functionality meets the beauty.

Aqua Sheko Soho

Title: Aqua Sheko Soho
Design Agency: 7Gods
Location: Frith Street, Soho, London
Client: UK Fish Spa Limited
Photography: Chris Rentos Photography

introduction:

After the successful launch of Aqua Sheko Kensington spa, 7Gods had been asked to repeat the success of its first concept but on a larger scale. The dark décor used to marge and keep the continuity of the first spa, some elements were changed to give its own character. Inspired by the vibrant atmosphere of Soho, using dark purple with white wall patterns, high gloss finish and LED lighting, passersby would be forgiven for thinking the sleek new split level store is a more bar than a spa. New to Aqua Sheko is the hand treatment bar on the ground floor. The downstairs level feels as it has been transported from another world, overlooking a tranquil Japanese inspired garden, which forms the center piece of the room. The basement level is also home to two full body tanks where the clients can submerge themselves in a pool of fish for all body exfoliation.

The Soho Aqua Sheko is held at the much bigger venue. It is a stunning visual masterpiece, where the fish tanks are originally adapted into the whole design experience, highlighted with the diffused blue light vibrantly. Use to sterilize the tanks, ultra violet color is being both a necessary functional purpose and a design concept of the tropical reef, where shiny, fluorescent fish tanks reflect the sun light beams through the water surface.

AQUA
SHEKO

1

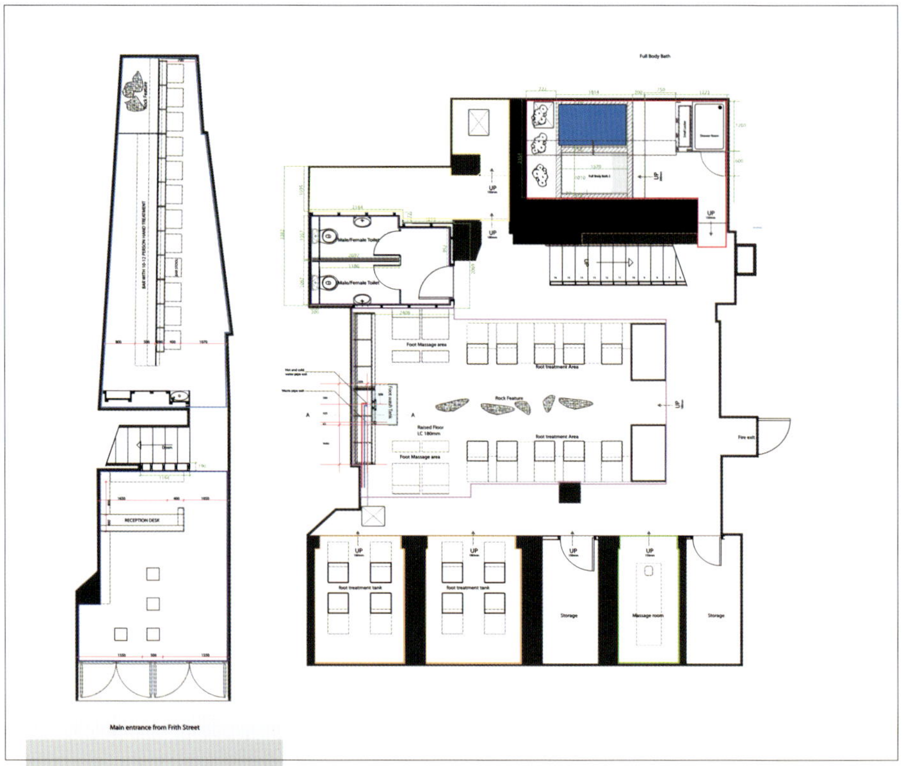
Full Body Bath
Main entrance from Frith Street

AQUA
SHEKO
IMPORT &

Mandi Bali Kobe

Title: Mandi Bali Kobe
Design Agency: Kamio Design, Ichii Corporation
Location: Kobe, Japan
Client: EBM Co., Ltd.
Area:600 m^2
Photography: Nacasa and partners

introduction:

Mandi Bali Kobe is a Japanese spa shop providing Balinese skincare-massage service. As we were expected to express a sense of Bali, which is loved by a lot of Japanese ladies, we also tried to find a new international design style through this project.

All over the world, there are millions of commercial spaces designed by millions of cultural styles such as European, American, Chinese, and Japanese. However, you may wonder how many shops actually express the internationalism through their design concepts. Surely even within Japan, many styles and ideas are enough established and distinguished to each other, such as simplicity, gorgeousness, contemporary, and classic, and those design qualities are never something more or less than one may found in western countries; it is obvious that Japan has enough high qualities of design in commercial fields. However, this does not mean Japanese design has become a world wide standard and Japanese designers always consider of such a strong internationalism, which may get over a typical sense of cultural backgrounds.

We should analyze and look for more on a common design sense to be understood and shared between many countries. No more only Japanese style. No more only Balinese style. We believe that establishing a new international design style may expand possibilities of designers not only for Japan but also for all Asian countries.

We imported all raw materials directly from Bali such as stones and tiles, and we expressed them in Japanese geometric theory. Also, by combining the design aspects carefully, rather than just an individual spa shop feeling, we tried to provide a sense of luxury hotel, which many people may visit from all over the world. Somewhere between Bali and Japan. Somewhere between West and East. Somewhere between Classic and Modern. We studied for an internationalism somewhere between all of us.

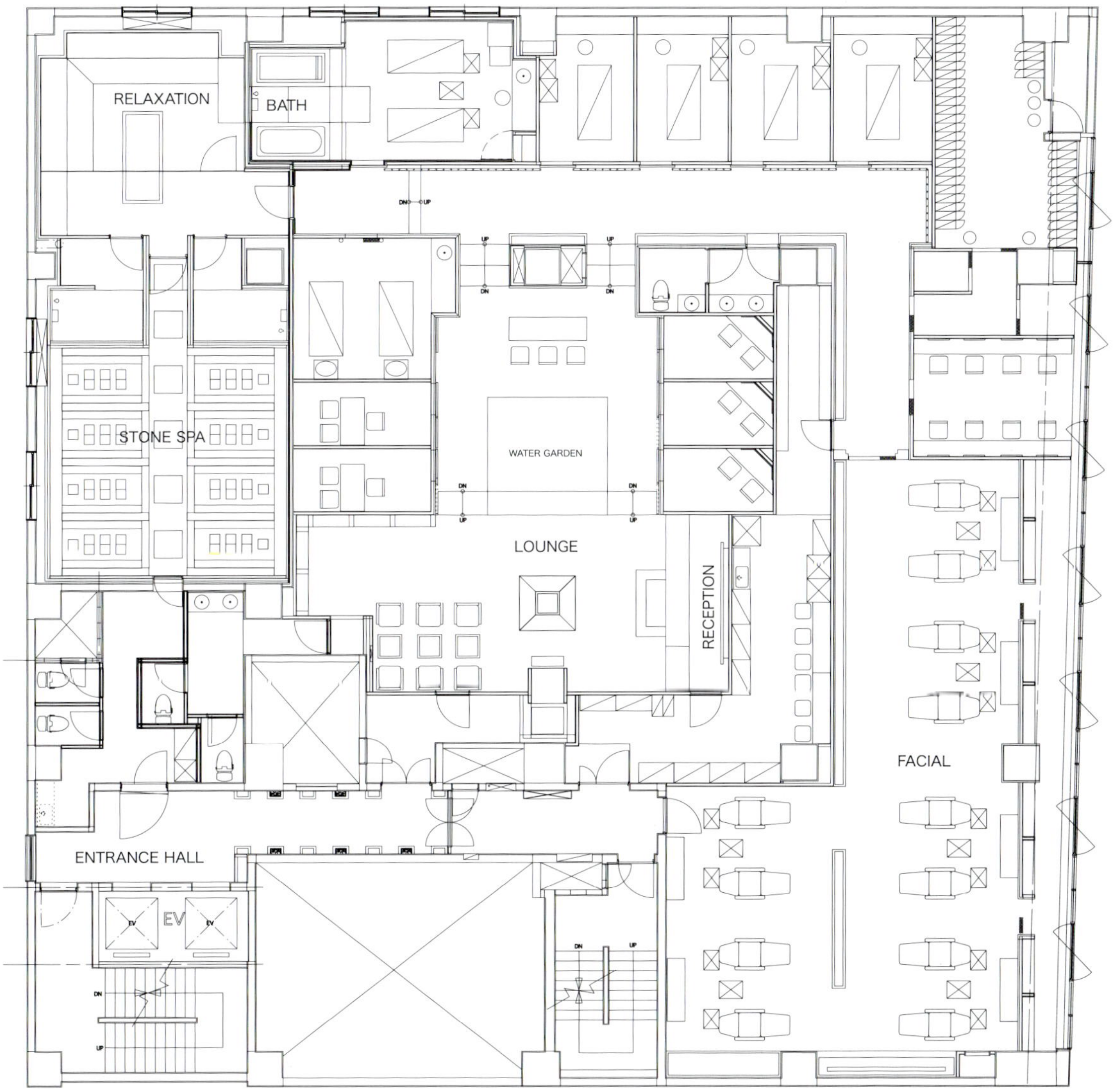

1

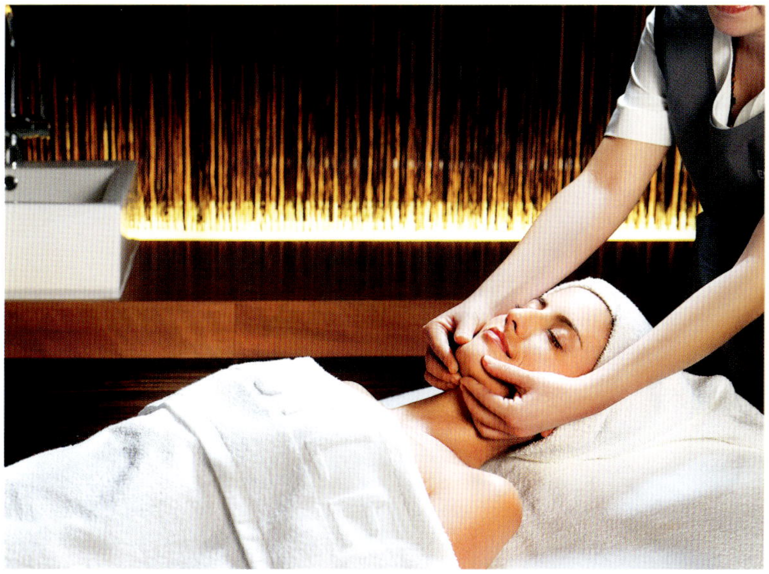

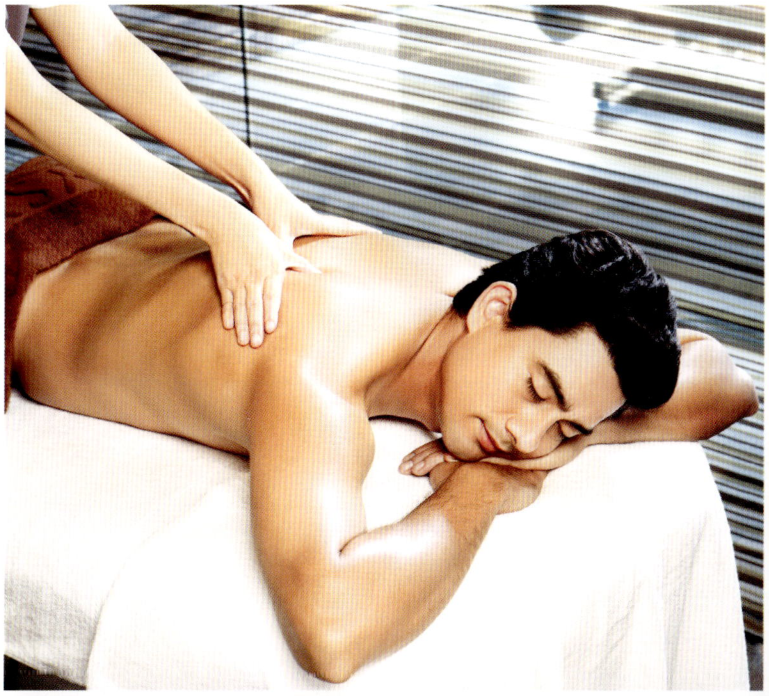

ELYSYLE
100% Pure & Natural Oil
Made in France
100ml
floral water
rosemary
carrier oil
wheat germ
restorative
ELYSYLE
Made in France
10ml
107
rose
geranium

ELYSYLE
100% Pure & Natural Oil
Made in France
100ml
restorative
100% Pure & Natural
100ml
RHU
ELYSYLE
100% Pure & Natural Oil
Made in France
10ml
ELYSYLE
100% Pure & Natural Oil

Scandinave Les Bains Vieux

Title: Scandinave Les Bains Vieux
Design Agency: Saucier + Perrotte architectes
Location: Montréal, Québec, Canada
Client: Gestion Rivière du Diable
Area:1000 m^2
Photography: Marc Cramer

introduction:

Located in the heart of historic Old Montréal and facing the vibrant piers of Old Port, Scandinave Les Bains Vieux is an urban spa whose purpose is to provide a thermal therapy experience that engages each of the body's senses. The building, rebuilt half a century ago after extensive fire damage, has housed warehouse functions until recently when it was acquired by the new owners: the new spa is positioned at the ground/entrance level.

The formal parti of the project is derived from the contact between hot and cold — and more specifically, the naturally occurring phenomena associated with these conditions — the design distills the idea of cool glacial forms and the warmness of volcanic rocks. White angular masses of glacial topography coupled with volcanic geology bespeak the duality that is central to the thermal therapeutic experience proposed by the spa. This duality is articulated through both the spaces' forms and the selection of various materials.

Upon exiting the dressing room, the visitor is immersed in a unique environment where walls, floors and ceiling are slightly angled according to a notion of interior topography through which visitors may wander. These angles, though subtle, give bathers a perceptual difference from their everyday environments; the awareness of the corporeal relationship with their surroundings is heightened, thus grounding each visitor for that moment in time. Just as in a natural landscape, slight undulations in the ground plane create gentle slopes; depressions in the floor level generate basins of water for bathing. At particular moments, volumes emerge from the ground to sculpt interior

zones for the sauna and steam bath. Uniting the main space is an undulating wood ceiling that echoes the movements of the floor: walls of white marble mosaic appear to melt at the point of contact with the warm-colored wood on the ceiling, resulting in accentuated architectural reveals. Heated, cantilevered benches made of black slate offer visitors a warm place to pause in between a hot and cold bathing cycle.

Opalescent glass has been added to admit natural light through the building's existing openings while providing a sense of privacy for the visitors. The light that permeates the bath area glows, adding to the purity of the space and the feeling of tranquility for the bathers while keeping contact with city life. Along de la Commune Street, a thin cascading layer of water flows on glass surfaces, filtering views so that from the exterior, passersby can see only shadowed silhouettes of the figures within the hot bath. Rounding out the holistic journey is a relaxation room where bathers can relax in rocking chairs or bean-bag lounge chairs.

1

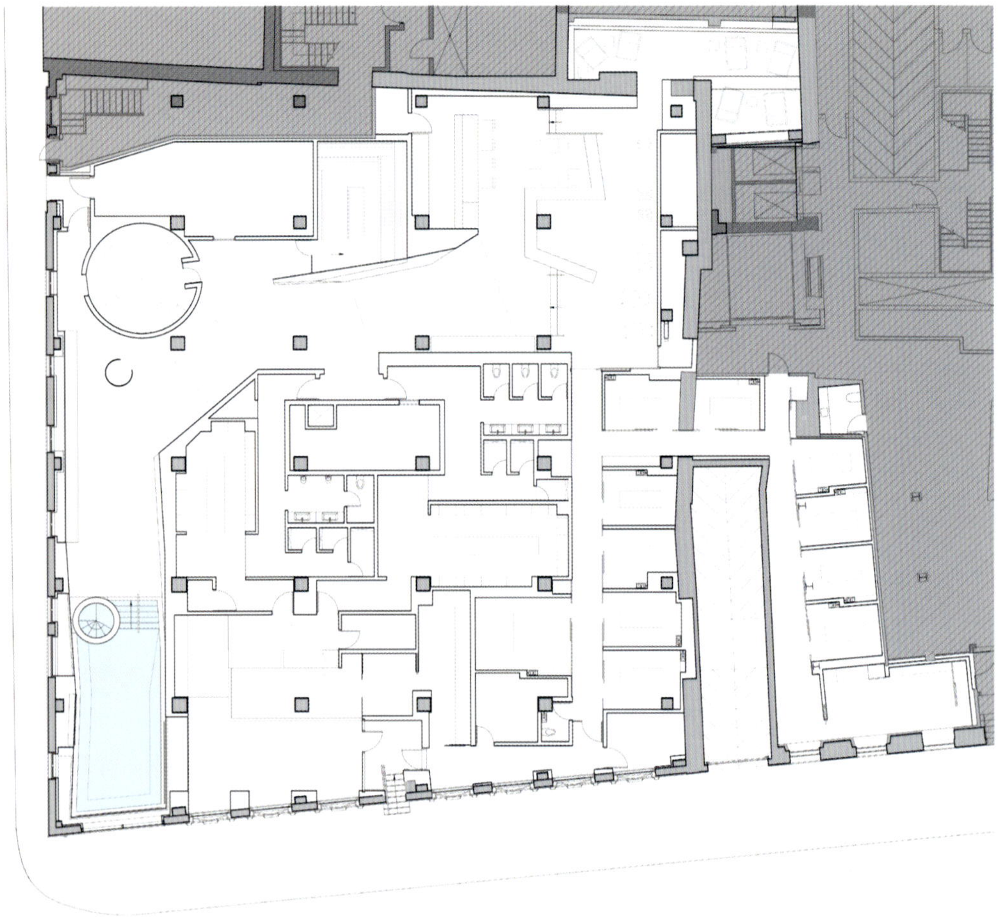

ego

Oasis Club

Title: Oasis Club
Design Agency: PAL Design Consultants Ltd
Location: Chengdu, China
Client: Townowner Real Estate
Area: 25,000 m²
Photography: Bao Sze Wang & Sharon Sze

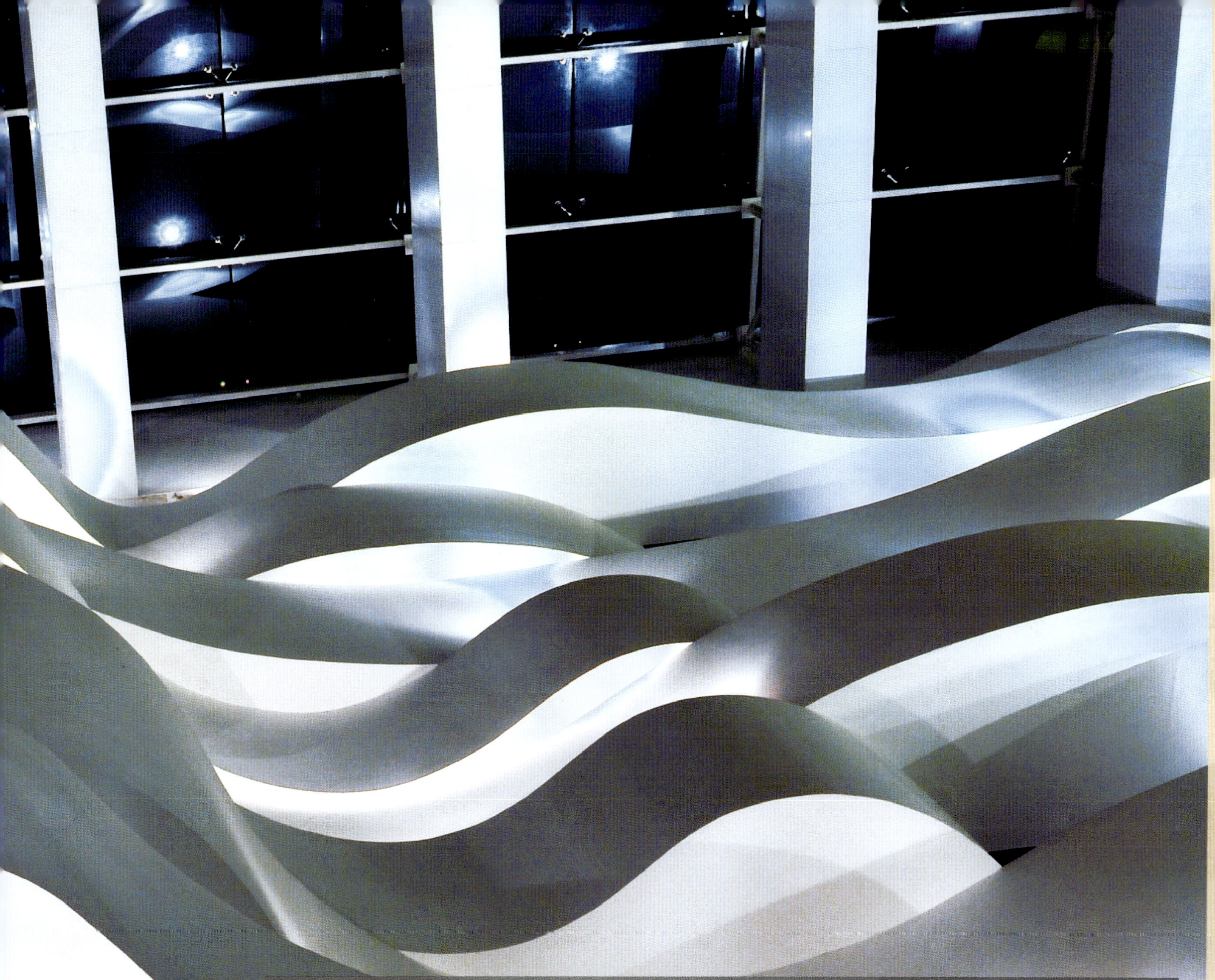

introduction:

With new concept of spatial design, Oasis Club is a chic private clubhouse of luxurious residential project located at the continental area Chengdu, China. With four floors of total 25,000 sqm for different functions, the design concept brings the mix of water, spa and funny elements transform as an oasis at the inland. A lot of wave curve walls presented irregular shape in the endless and motive spiritualization. The theme color palette consists of navy blue, lake green, pure white for main area, plus the party color of orange, yellow, and purple for young clients in the peace and calm environment for leisure.

It brings ocean theme at the entrance zone providing cafeteria services. A main reception with lake green wall panel is located at the entrance of first floor. Besides, wave stairs walked towards café. Turn upward to the roof top, a shock of 3D wave powerful curve at four floorings feature wall is exhibited. There is a set of contemporary urchin-like lighting at the atrium become a sparkle sculpture to radiate the life in the ocean. Those treatment rooms with navy blue curve walls provide popular foot massage services for the residents.

A complex of spa facilities is completely fitted for all residents in the common floor. The color scheme turned to light tone. At the reception of spa, purple dots are perfectly decorated with bright white arc-curved feature wall to express its clean space. An icon feature is displayed at the main entrance for creating a sense of future. The pathway leads residents to explore the core of shell. The first adventure is a central locker area designed with black and white locker boxes. Besides, there is an oval-shaped mini swimming pool for public. Face to the pool, a row of showers, sauna room and steam room keep more convenient and user-friendly. Behind the center point, there is a lounge with extreme illusion space design. Bright cheerful colors such as yellow, orange, blue have indicated the party dynamic.

Through curved channel beyond the magic imagination towards the magazine corner which is a long path with chic and comfortable sofa for friend gathering.

Come to the treatment zone, space design as a labyrinth, it provides 16 spa treatment rooms for soul and body juvenescence. This level turned to the dark scheme such as jade green, purple blue and golden honey for soft lighting effect. The top floor is VIP spa zone with 8 luxurious spa treatment rooms for exclusive residents only. Each private room provides a mini theatre, Jacuzzi, sauna and steam room. Some of them include a balcony for outdoor Jacuzzi and sun deck. The biggest one provides a small indoor pool with temperature control. With the décor of mosaic shell and pebble stones, it aims to create a private relaxation in the paradise, dreaming into a sea palace of the myth.

1

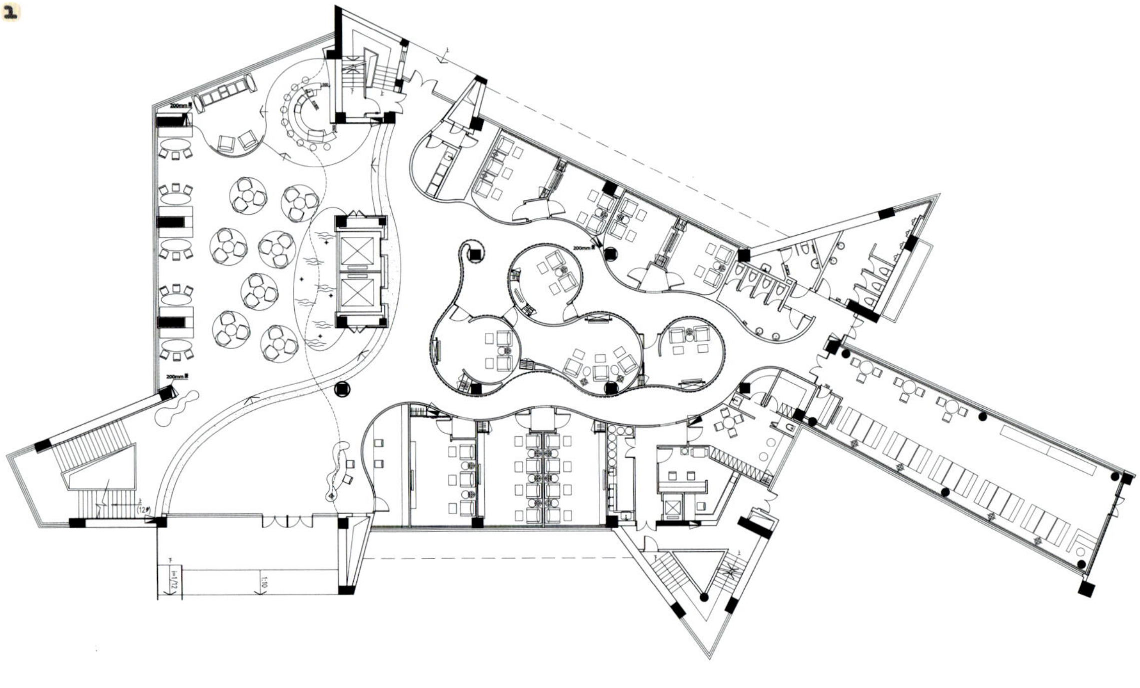

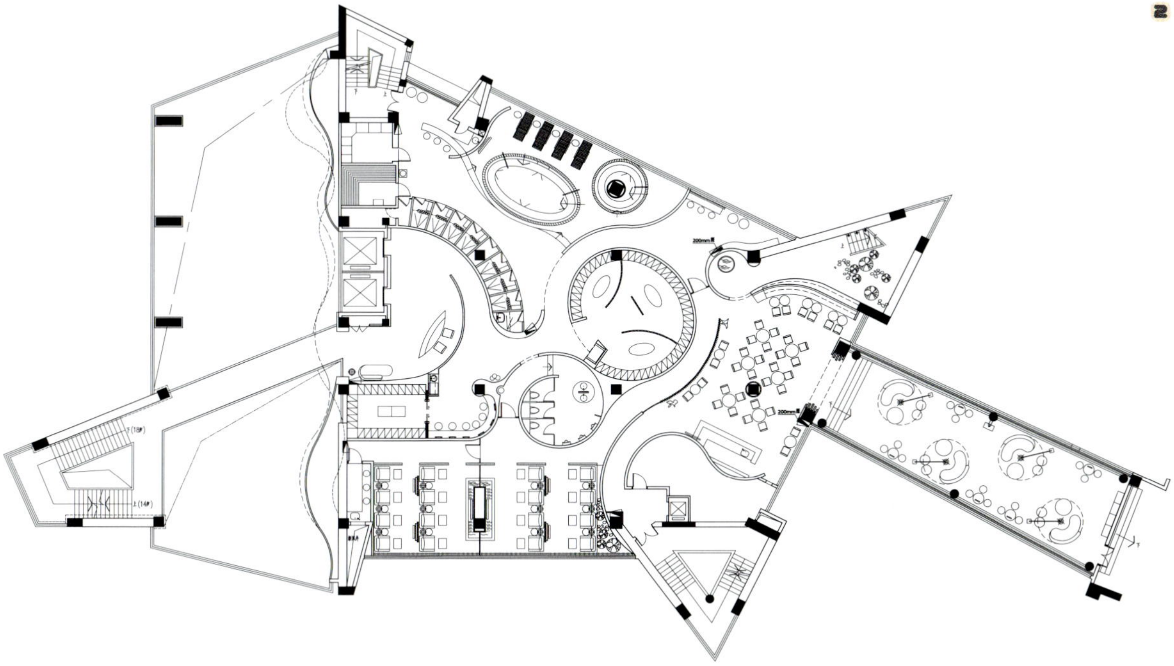

2

3

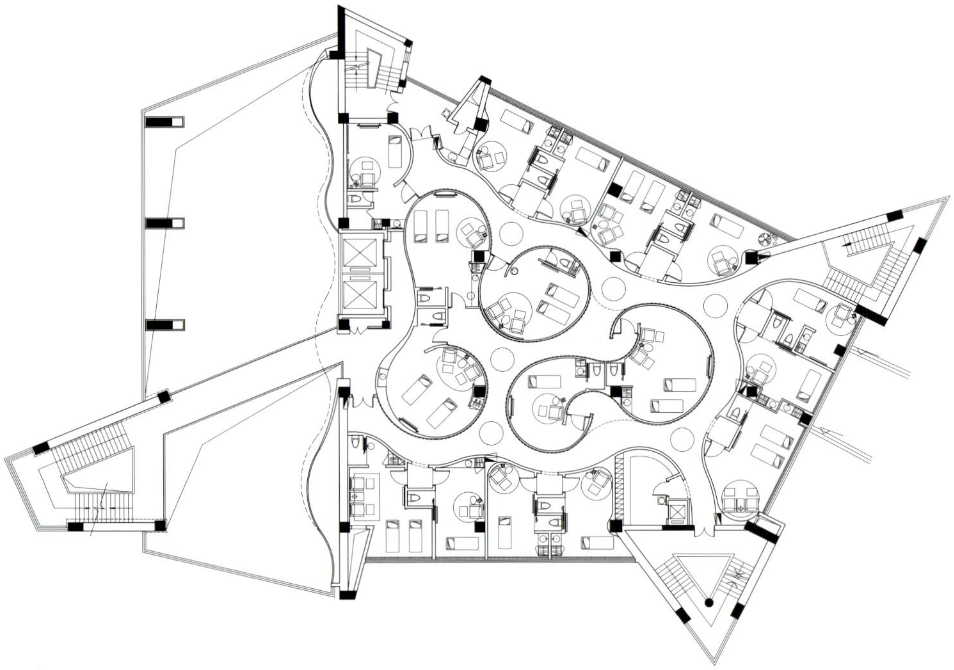

4

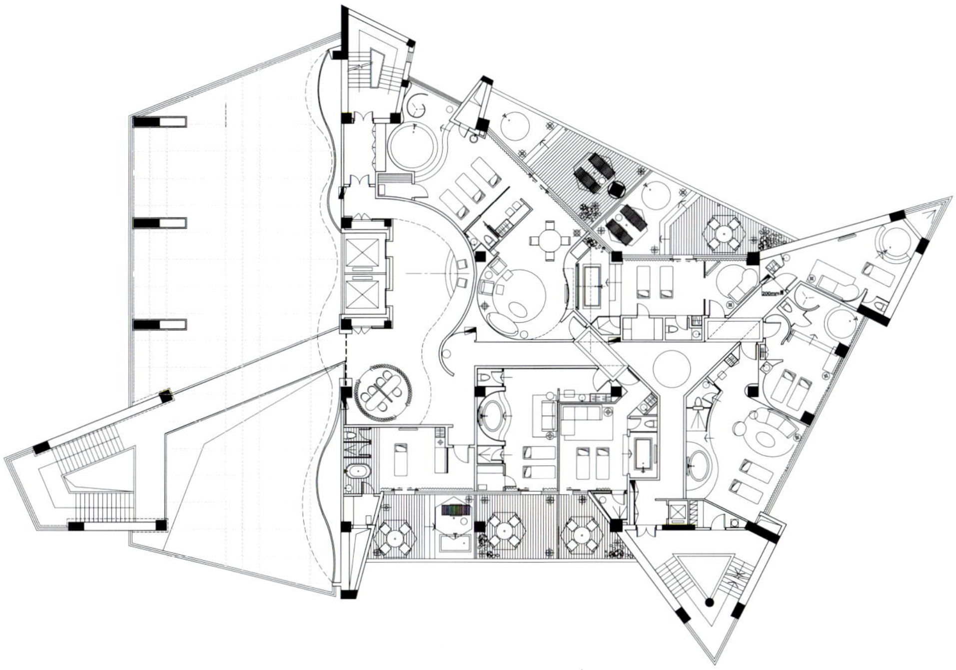

The Setai Club at the Setai New York

Title: The Setai Club at the Setai New York
Design Agency: Robert D. Henry Architects (RDHA)
Location: New York, USA
Client: The Setai Group
Area:1115 m^2
Photography: Daniel Aubry Photography

introduction:

BREATHE | The spa offers a sensuous retreat from bustling New York City. A long latticed gallery greets you upon arrival; its rhythmic bays begin to pace one's breath within the expansive 12,000 square feet environment. A warm color palette of ivory, gold and chocolate with tinged red accents, creates a visually calming atmosphere. Subtle illumination emanating from behind the reception desk beckons guests to begin their journey, carrying them through the experience.

UNWIND | Reflections of light form diamond patterns on the inclined pathway leading to the Aqua Grotto within each of the changing areas. A raised marble plinth, contains a hydro bath, accentuated by trickling water on golden glass tiles, falling into the pool below. The sound of the hushed water, the visual dancing refracted light from the pool, combined with delicate aromatic scents and the warmth of the pools completely engage all the human senses. A steam room and sauna round out the thermal circuit, while guests unwind in chaise loungers. Softly lit hallways guide one through the treatment areas beginning with an elaborately screened tea lounge. Patrons slip into serenity while sipping exotic teas and melting into plush oversized lounge chairs. Two unique Experiential Couples Suites contain deep soaking tubs surrounded by mint green glass tiles. Luxurious showers and massage tables provide the chance to relax, integrating the soothing effect of the water. Pocket doors divide the suite into two treatment rooms for additional privacy during exclusive treatments. Six multi service Care Suites are enveloped floor to ceiling with elegant leather panels, rich dark walnut wood complimented with an ivory glass mosaic accents within a cocooning ivory envelope.

ENERGIZE | Beyond the treatment area resides a fitness and movement studio with walnut framed mirrors and millwork bring a stylish presence to a yoga or Pilates session. A full glass entry wall provides an unobstructed view to the high-tech fitness equipment against a back-drop of animated fiery red. The 'high energy' of the fitness studio purposely contrasts with the 'calming energy' of the spa.

LEGEND

1 SPA RECEPTION
2 WOMEN'S SPA ATTENDANT
3 WOMEN'S LOCKERS
4 WOMEN'S SHOWERS
5 CO-ED LOUNGE
6 WET TREATMENT ROOMS
7 DRY TREATMENT ROOMS
8 YOGA ROOM
9 MEN'S LOCKERS
10 MEN'S SPA ATTENDANT
11 MEN'S WATER SUITE
12 WOMEN'S WATER SUITE

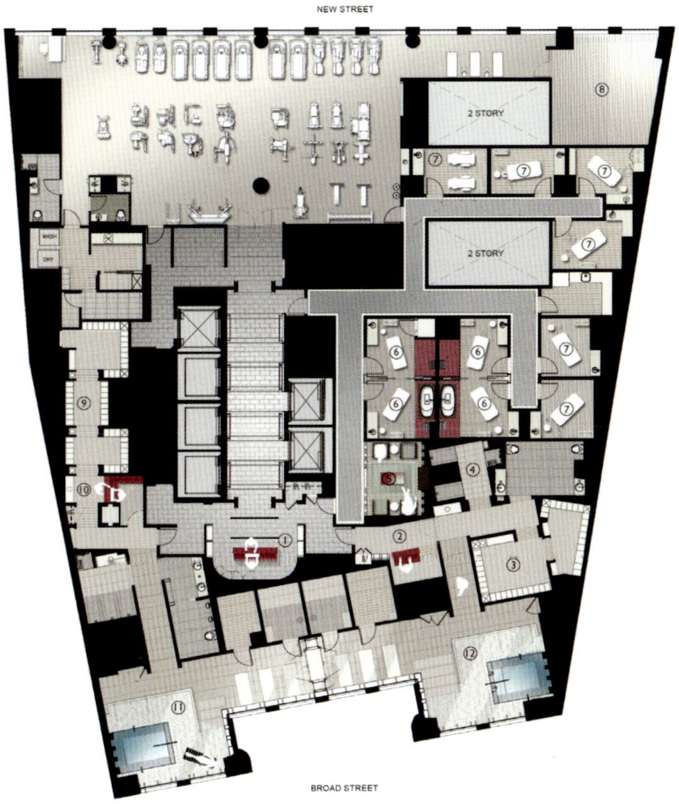

1

HOME SPA

Title: Home Spa
Design Agency: Spaplan
Area:60 m²

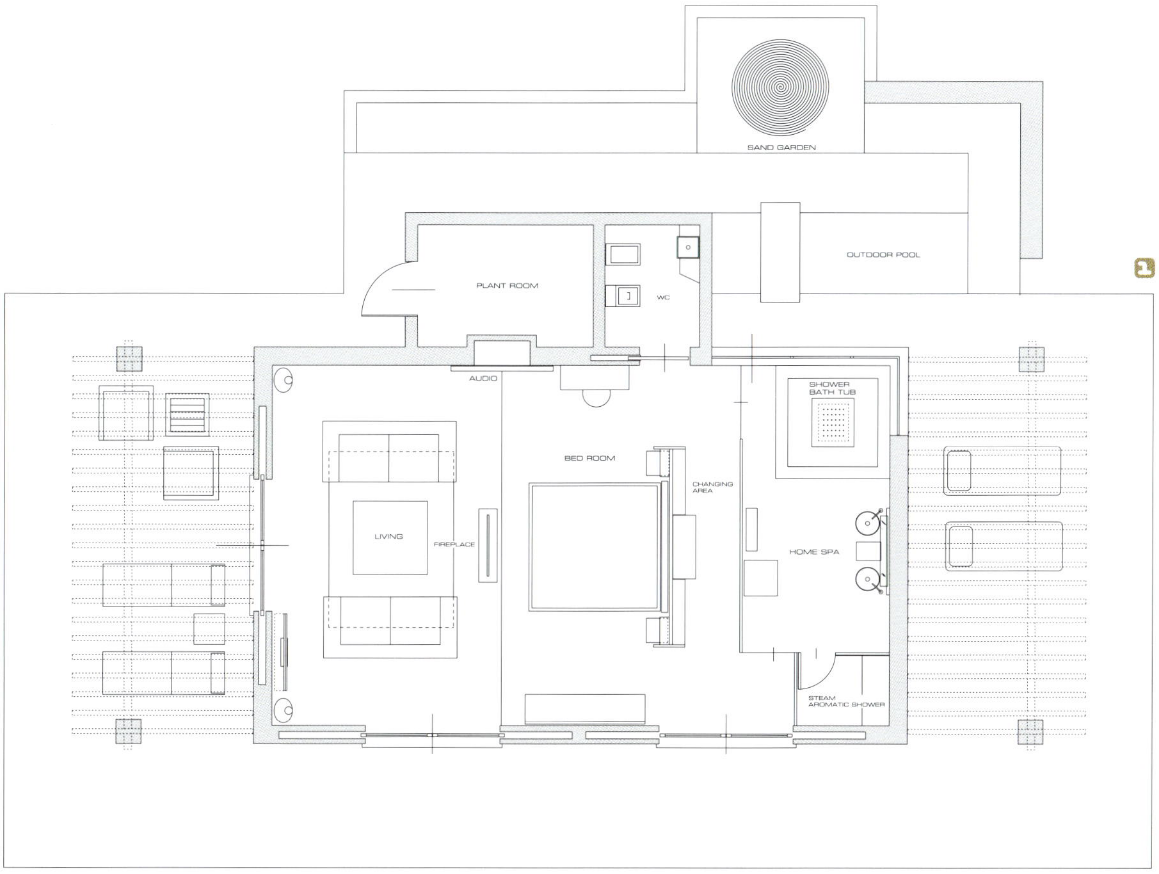

Introduction:

In the heart of peaceful Tuscany countryside surrounded by olive trees on sweet hills, this private suite room is not only an oasis far from the stressful metropolitan life but gifted by generous nature the abundant hot spring originated from Roman epoch.
A private suite room is a small and cosy house realized inside a spa resort property. It permits enjoying all comfortable hotel services at the same time keeping complete privacy around the house.

The external building is totally in white color combined with natural wood grating shutters. Home spa with bath room is visible from every angle through transparent wide glass wall.
All house design is inspired from typical Tuscany nature, simple, homely and natural which brings us in harmony with Nature's equilibrium.

Essential Home Spa combined with elegant bathroom and organized by steam room, aromatherapy, ceiling shower with various relaxing and massage functions like nebulize shower, sequenced waterfall and rain showers etc.
Small outside pool is next to Trani white marble bath tub separated by transparent glass wall.
After sunset, the outside pool is lighted up in blue naturally by photoluminescence stones charged by sun light during the daytime.

Today in Italy Home Spa is making a boom by the people who are more sensitive to maintain and control his own health relieved from everyday stress.

Manzi Natural Spa

Title: Manzi Natural Spa
Design Agency: Spaplan
Location: Ischia Island, Italy
Client: Terme Manzi Hotel & Spa
Area:1600 m^2

introduction:

Ischia is volcanic island with generous rich mineral hot water rises everywhere in the island.
Terme Manzi Spa was built in 1863 on the historical spring called 'Gurgitiello' by the Ancient Roman people who loved this island as the excellent spa resort 2000 years ago.

The most famous Italian General Giuseppe Garibaldi who fought for country's independence stayed at Terme Manzi to heal war injury.
The new spa project paying all attention to important building history and local color faced courageously to propose ultimate spa concept. Manzi Natural Spa planned for this luxurious 5 star hotel which gives hospitality often to discerning international guests.

Spa concept is focused on the nature with which we are a part of it.
Feeling be natural and relaxed.
1600m^2 spa is an silent oasis spreads on two wide floors. Numerous wellness courses are planned in organic forms starting with a curious visit on historical bath tub used by General Garibaldi followed by eccentric style reception,esthetic and mud treatment rooms, hydrotherapy, shinny and particular round relax area, the Roman bath and inner pool on the ground floor, suit couple treatment rooms , multi-treatment room and fitness on the superior floor.
The design is simple and only authentic natural materials are selected accurately to match comfortable relaxing atmosphere.
Natural wood is the key design in the spa. For example wood floor is certified as 300 years old oak recovered from old houses and used again. The round spacious relax area is accented by rhythmical and sculptural wood lamps making a soft and gentle mood.

1

INHALATION THERAPY
SHIATSU MASSAGE
JET MASSAGE
MULTI TREATMENT
4 HANDS SHOWER MASSAGE
UNDERWATER MASSAGE
MULTIJET
GARIBALDI'S BATH TUB
ENTRANCE
SELF-TREATMENT
SHOWERS
TEPIDARIUM RELAX
OFFICE
RECEPTION
SOLIUM
SOAP MASSAGE
SPA MANAGER DOCTORS OFFICE
LACONICUM
SUDATIO
PLANT ROOM
WC
HEALTH BAR
BEAUTY TREATMENT
WAITING AREA
BEAUTY TREATMENT
WC
WC
WC
BEAUTY TREATMENT
BEAUTY - MUD TREATMENT
MULTI-TREATMENT
MUD TREATMENT
MUD TREATMENT
RELAX
STEAM
INDOOR POOL
MASSAGE- RELAX
WAITING AREA
RELAX
RELAX
PHYSIOTHERAPY MASSAGE
MUD TREATMENT
MUD TREATMENT
MUD TREATMENT
MULTI - TREATMENT
STORE
STAFF ROOM
MULTI - TREATMENT
MULTI - TREATMENT
STAFF ROOM
STAFF ROOM STORE
CHANGING ROOM
WC
CHANGING ROOM

REDUCE DA ASPROMONTE
GIUSEPPE GARIBALDI
IN QUESTO CAMERINO FECE IL BAGNO

REDUCE DA ASPROMONTE
GIUSEPPE GARIBALDI
IN QUESTO CAMERINO FECE IL BAGNO

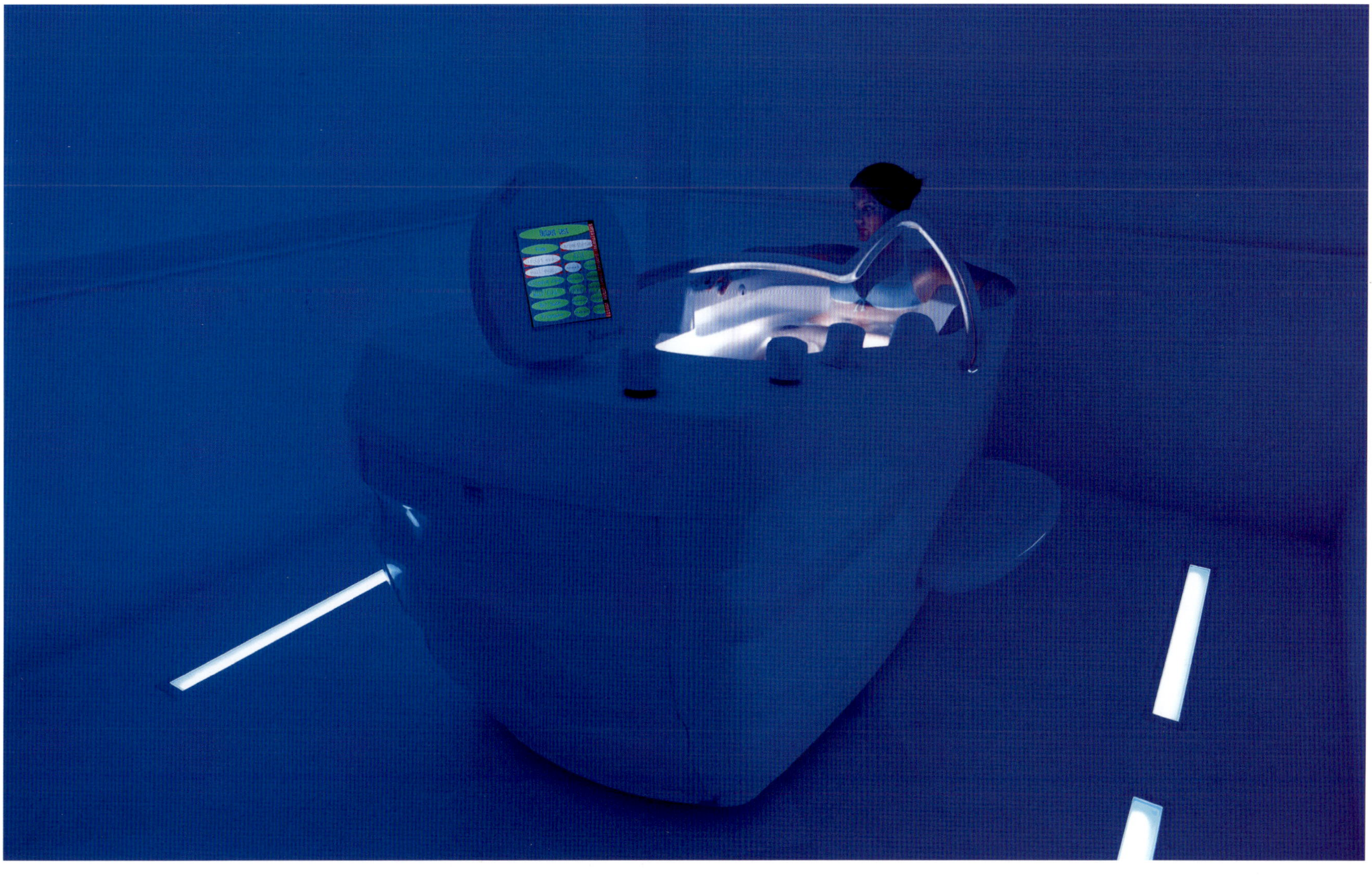

Kanebo Sensai Select Spa

Title: Kanebo Sensai Select Spa
Design Agency: Curiosity
Location: Interlaken, Switzerland
Client: Kanebo Cosmetics
Area:242 m^2
Photography: Nacasa & Partners

Introduction:

The Sensai experience is a journey through sequences of sensorial encounters. The Japanese esthetic is a subtle balance of simplicity created from complexity.

The design is based on koishimaru silk, the core value of Sensai brand. The space is lined with floating layers of silk. The soft glow of light gently removes layer by layer, and the outside world transforms mind and body. The waiting room in the center of the space is designed as a 'cocoon'—an oasis of tranquility surrounded by serene light. The everyday life gives way to a different dimensions, leading to an unexplored sense of relaxation and well being.

The darkness of the corridor contrasts with the soft glow of the cabin light,inspired by the simplicity and reverence to nature. A lighting wall is created with layers interlaced strings of fabric, as the eyes loses focus, the space vanishes and the mind liberated. Reflection, refraction, distortion, light and shadow create a play for the senses. Space, material and light fuse into immateriality, where body and mind turn into harmony.

1

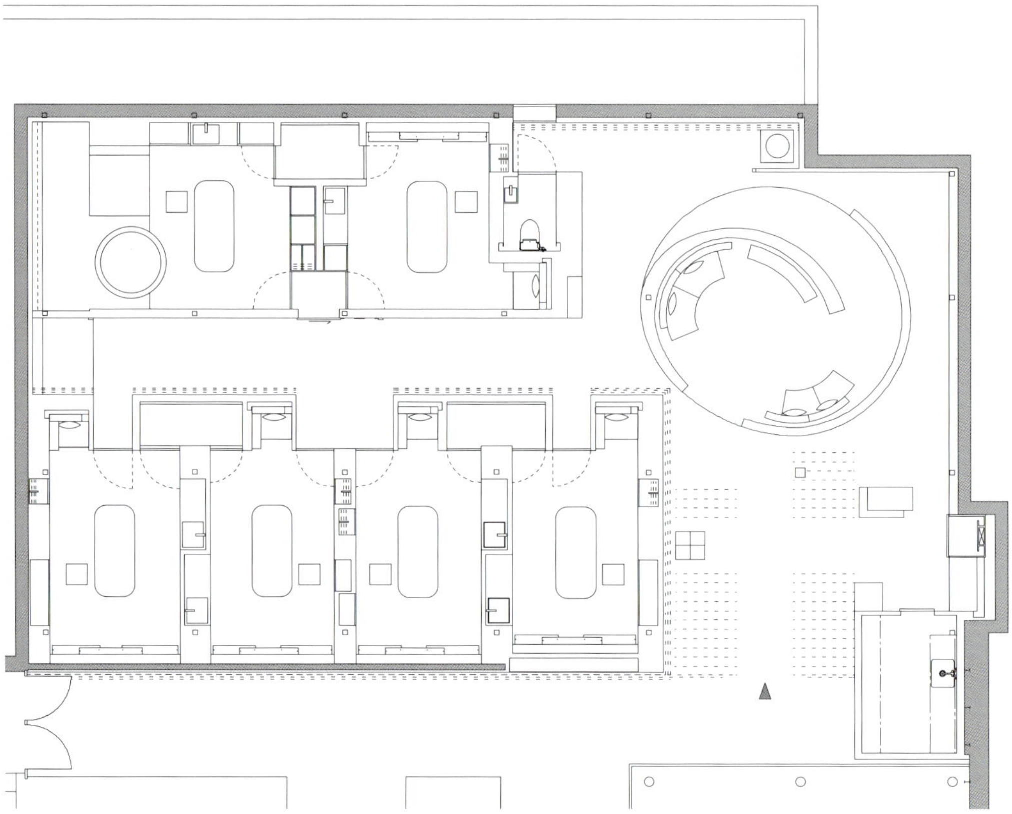

Mandi Bali Nagoya

Title: Mandi Bali Nagoya
Design Agency: Kamio Design, Ichii Corporation
Location: Nagoya, Japan
Client: EBM Co., Ltd.
Area:335 m^2
Photography: Nacasa and partners

introduction:

This is the second project for MANDI BALI style spa. While we focused on and studied the new internationalism design in the previous project, this time we concentrated more on how people may find the same ambition as in 'Mandi Kobe' even though actually it was much smaller area. Honestly, it was true that the advantage of largeness could be the one of strong aspects to make the project successful. Without this advantage, we needed to deal with the limited space to provide the luxury and international feelings, which may be found in some kind of high classed hotels all over the world.

Rather than taking many design aspects, we tried to pick up only one theme to express a high-end space; we decided and designed the spa lounge as a special private library. If we see the waiting space just as a lounge, it could be a bit tight. But, if we consider it as a private library, it could be very special and may be enough spacious to make a luxury ambition.

We placed the reception and actual waiting space to the center of the shop area and surrounded it by the library-like corridors with indirect lighting effects. For the partition between this center space and the corridors, we applied clear glass walls so that the guest may see through the corridor-library and may feel deepness around them.

It is always designers' nightmare to deal with a limited space. However, by taking the corridors as a part of main aspect and giving them a strong character, we figured out an answer, which may be very worth to both of the shop owner and guests.

MANDI-BALI

Sayan

Nagomi

Title: Nagomi
Design Agency: Kamio Design, Ichii Corporation
Location: Matsudo, Japan
Client: EBM Co., Ltd.
Area:300 m^2
Photography: Nacasa and partners

Hilton Mexico City Reforma Hotel Spa & Fitness Center

Title: Hilton Mexico City Reforma Hotel Spa & Fitness Center
Design Agency: Pascal Arquitectos
Location: Mexico
Client: Inmobiliaria Interpres
Area: 1000 m^2
Photography: Sófocles Hernández

introduction:

On the site where for decades stood the Hotel del Prado—which suffered damages during the1985 earthquake and was later demolished—the construction of an office building was planned. Later, given market conditions at the time, it was decided to build a modern design hotel that integrated the latest technologies and would have the possibility of adapting to new developments in coming years.

The idea was to break intentionally with its surrounding context and the image of its neighboring buildings, but that would, at the same time, manage to bring back the atmosphere and the commercial, cultural and social activities of Mexico´s 50´s, where hotels were considered socio-cultural gathering centers, such as the Hotel del Prado used to be.

In order to extend the business traveler´s stay over the weekend, a Fitness Center & Spa, and 3,000 sqf garden were included. This gardened area matches in a certain way with the original design of the old Alameda, but with a more contemporary landscape design; it overlooks the Alameda Park, surrounds the cafeteria, the lap pool and paddle tennis court. The spa services include massage and beauty treatment salons, a gym, an indoor swimming pool and social events open area. All areas are decked in teak wood and they are secluded from the main street but at the same time they participate of the urban context.

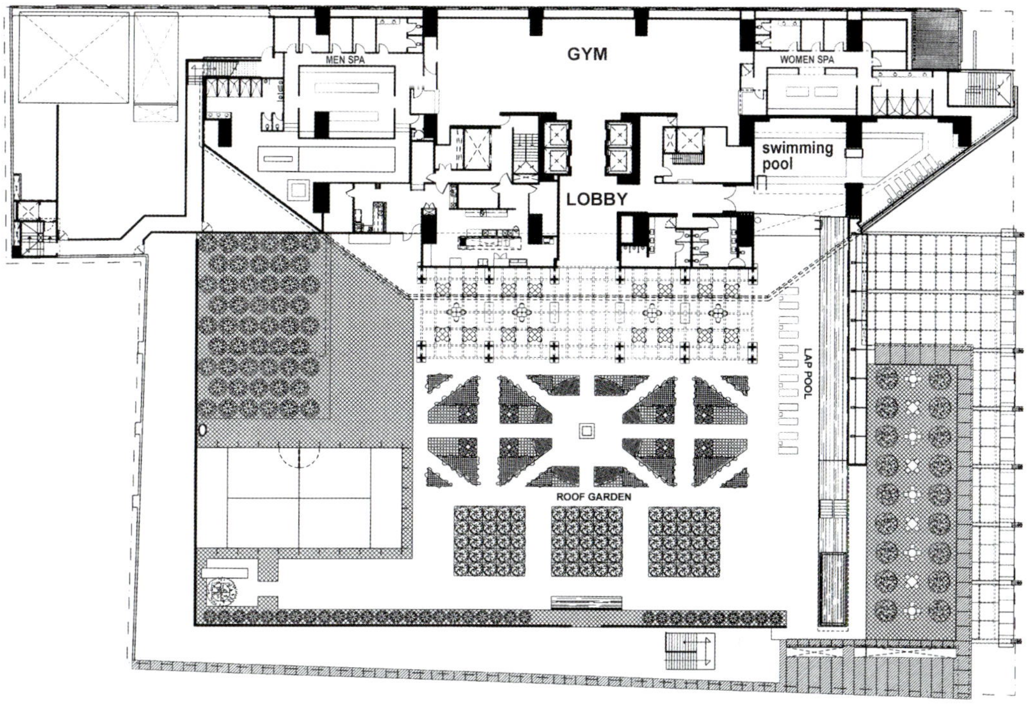
GYM
MEN SPA
WOMEN SPA
swimming pool
LOBBY
LAP POOL
ROOF GARDEN

LA CORTESIA EN EL DEPORTE

PRECOR
CYBEX
CYBEX

PRECOR

388
399
400
411
412

Banyan Tree Spa Shanghai

Title: Banyan Tree Spa Shanghai
Design Agency: Architrave Design & Planning
Location: Shanghai
Client: Banyan Tree Spa
Area: 1000 m²

introduction:

Located on the third level of The Westin Bund Center Shanghai, Banyan Tree Spa Shanghai spans 1,000 square metres, accommodating a total of 13 luxurious treatment rooms. This 21st century spa is specially designed by Banyan Tree's in-house design division, Architrave Design & Planning. The interior design of Banyan Tree Spa Shanghai displays a post-modern Chinese influence where the Five Elements are translated into a variety of refreshing ideas. There is a Fire Room created in flaming red, an Earth Room furnished in raw stones and a Water Room completed with a waterfall feature. The Yin Yang Deluxe Suite is the ultimate spa-goer's dream as it incorporates rain mist shower and steam facilities, bathtub with waterfall feature, a spacious relaxation area, two massage beds, changing area and a private toilet.

introduction:

There are 13 vip rooms with different themes, which integrate a variety of exotic delights. The exclusive famous salon for men comes up with the visual test for the designer. The mix and match approach to materials is the main axis through the design. And the 'concept' is the most important element of the case. Although there is no noble stone, the lines, color and geometrical figures are blended to produce an unexpected effect. Several simple elements of materials combined into different design patterns create the different space with sentiment.

The Japanese style full of chanverses is integrated with the circular wallpaper window grilles to create the atmosphere through the light and shadow. Which makes mystery exist tenderness, or the neon tubes filled with a sense of speed show the charming allure, or bold and avant-garde lips in strong red leave people a number of ambiguous reveries in a variety of compartments. The conceptual visual effects mix with all the basic elements of design to make people see distinctive design.

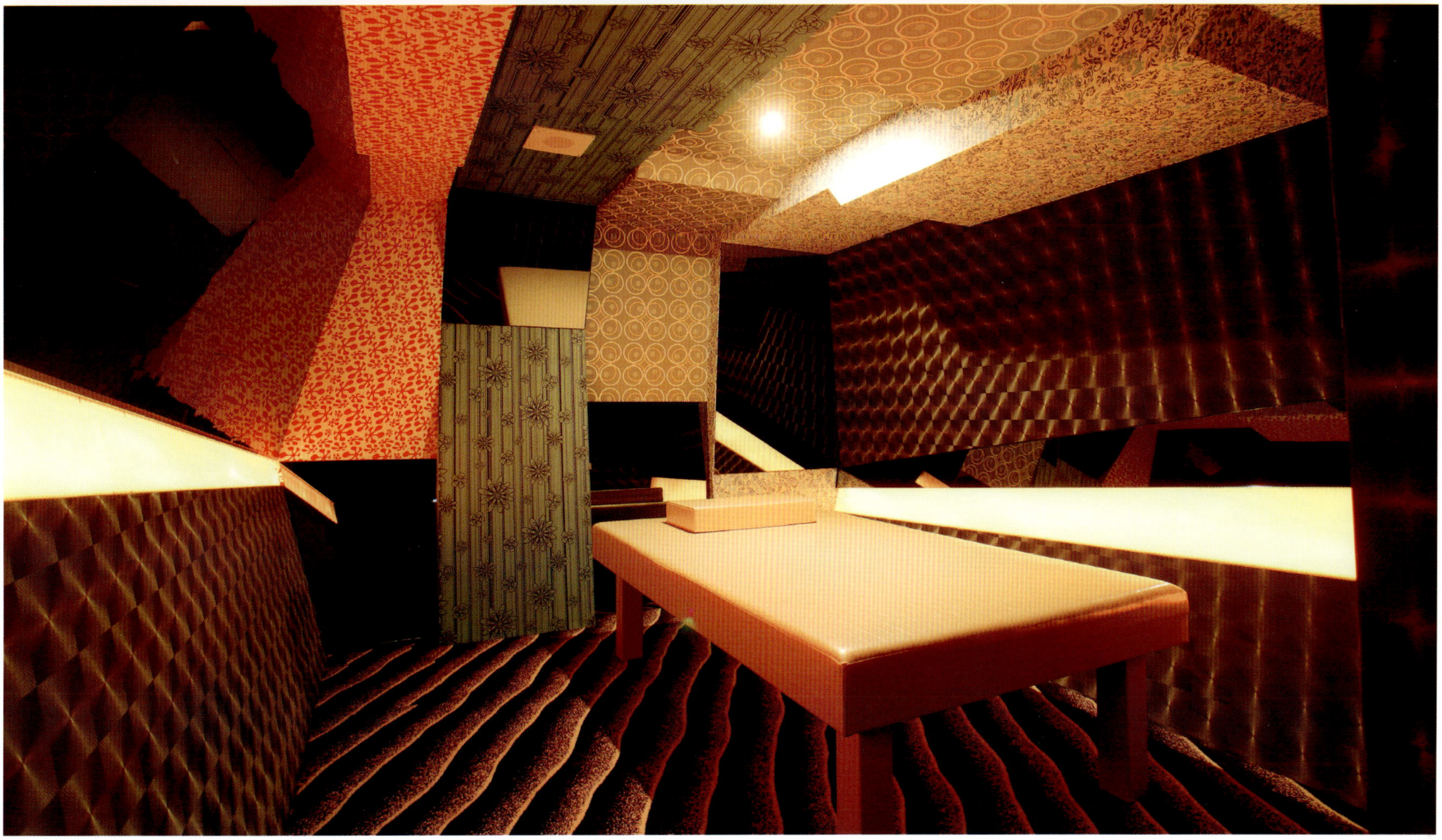

Art Peté

Title: Art Peté
Design Agency: Studio Alberto Apostoli
Location: Brescia, Italy
Client: Valentina Petenzi
Area: 116 m^2
Photography: Luca Morandini

introduction:

Art Peté is the result of a new conception of wellness, given by the architect and immediately shared by the ownership of the center.
The basic idea was to create an environment where the host, as well as finding spa treatments, could share with other clients some convivial time, ensuring in this way, his 'loyalty'.

From the architectural point of view, the concept is born from the symbolic central garden where everything comes to life and where there is a cabinet-sculpture used by guests to enjoy teas and fruit juices.

Apostoli used a lot of graphics, applied to glass walls, showing natural and relaxing subjects. A slatted wooden wall gives life to the reception, the cabinets for the products and merchandise, and goes through the whole center, interrupted only by the flush doors of the cabins which, when closed, disappear in the wall, creating a strong emotional impact. Inside, a simple and sober design communicates the idea of hygiene and freshness, and traditional materials as the wood and glass lend a sense of security.

A rational structure characterizes the geometry of the center where the form of false ceiling matches the design of ceramic floor. The soft lighting, created by veils, softens the bright green and beige walls.

2

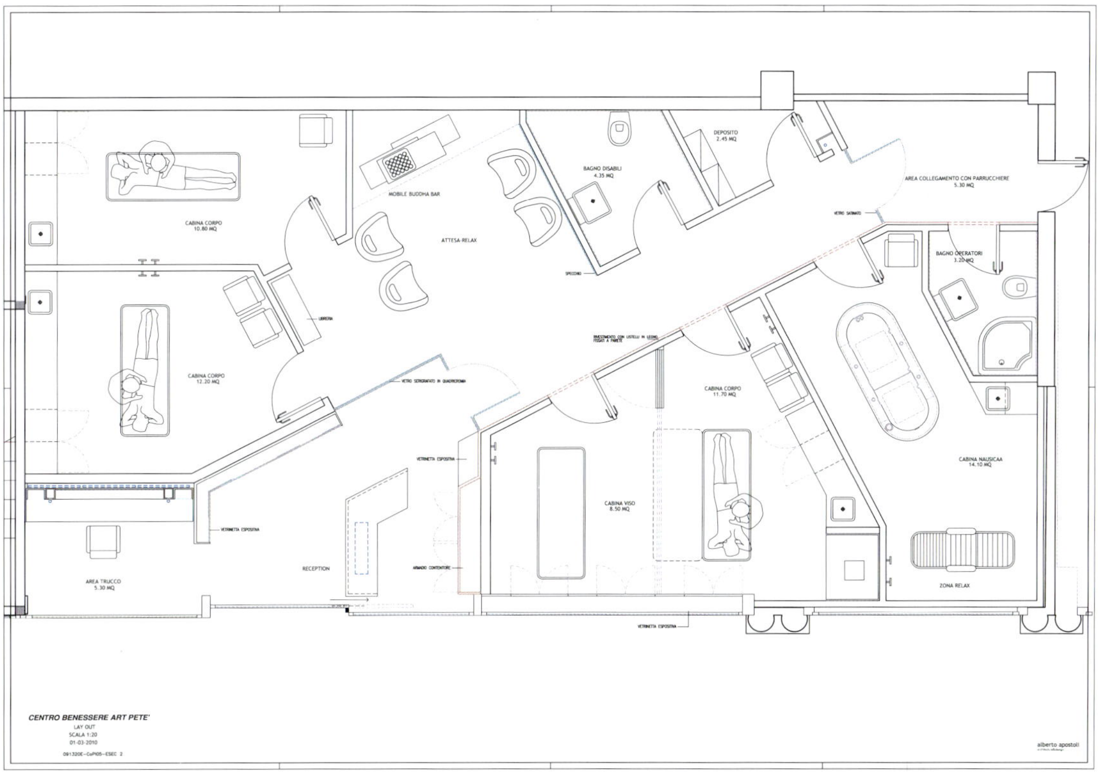

CABINA CORPO
10.80 MQ
MOBILE BUDDHA BAR
ATTESA-RELAX
LIBRERIA
CABINA CORPO
12.20 MQ
VETRO SERIGRAFATO IN QUADRICROMIA
VETRINETTA ESPOSITIVA
VETRINETTA ESPOSITIVA
RECEPTION
ARMADIO CONTENITORE
AREA TRUCCO
5.30 MQ
BAGNO DISABILI
4.35 MQ
DEPOSITO
2.45 MQ
SPECCHIO
AREA COLLEGAMENTO CON PARRUCCHIERE
5.30 MQ
VETRO SATINATO
BAGNO OPERATORI
3.20 MQ
RIVESTIMENTO CON LISTELLI IN LEGNO FISSATI A PARETE
CABINA CORPO
11.70 MQ
CABINA NAUSICAA
14.10 MQ
CABINA VISO
8.50 MQ
ZONA RELAX
VETRINETTA ESPOSITIVA
CENTRO BENESSERE ART PETE'
LAY OUT
SCALA 1:20
01-03-2010
091320E-CoPI05-ESEC 2
alberto apostoli

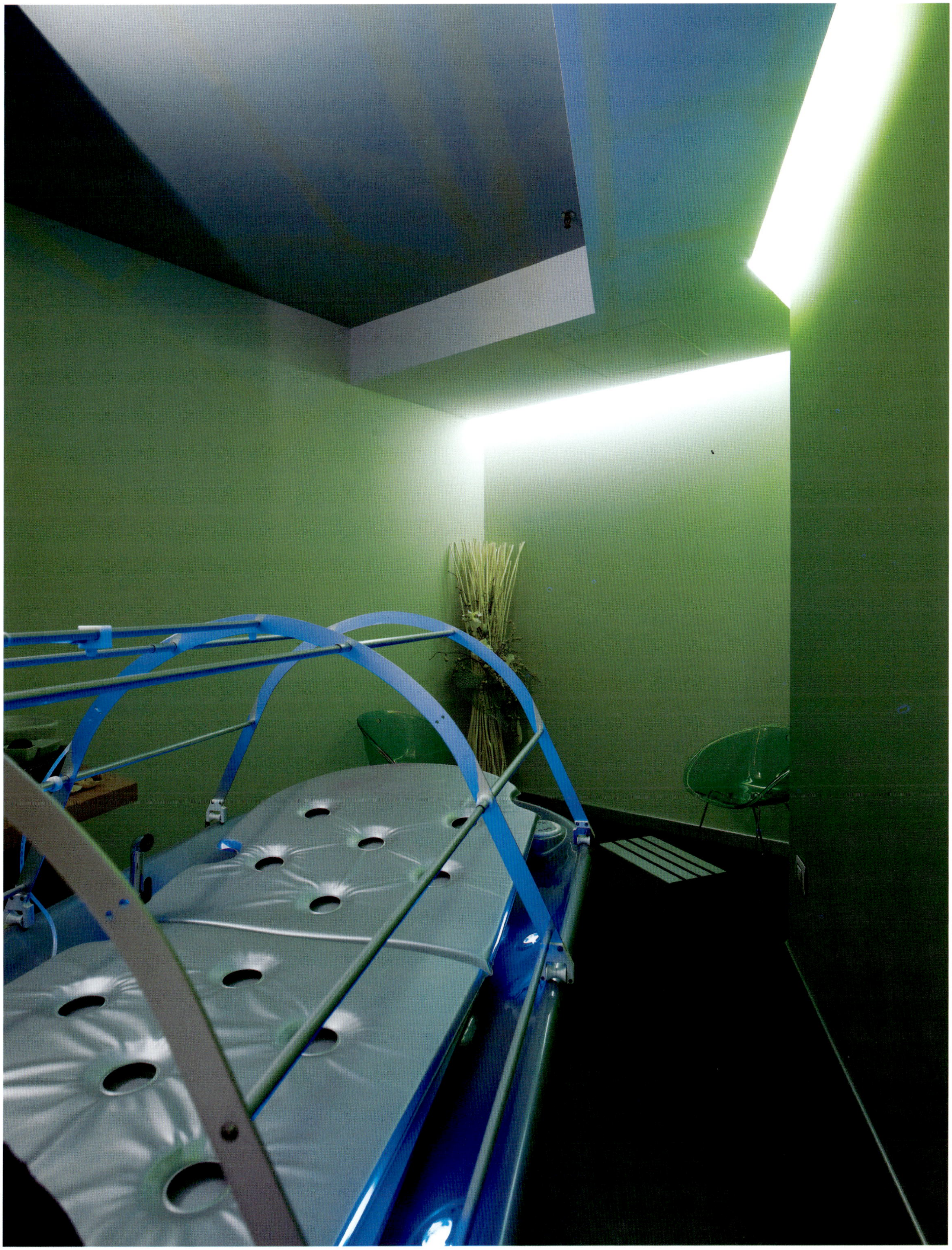

Sasha

Title: Sasha
Design Agency: Studio Alberto Apostoli
Client: Jacuzzi Europe spa
Area:8 m^2

introduction:

Made of wood, glass, steel and resin, this contemporary home spa is designed to recover the pleasure of self care and shape the most intimate essence of wellness. Sasha home is not only a full service of relaxing package but also stylish modern furniture to enrich your interior decor. Jacuzzi allows you to set the water, heat, music based on your need and this home spa is even completed with RGB LED inside the bench and the wall panel of the sauna space.The shower is surrounded by large glass panels and comes with three relaxing water programs to wash your cares away. Finally, the hammam coats your body with warm dew, while the touch screen panel allows color to intermingle with the different preloaded programs.

1

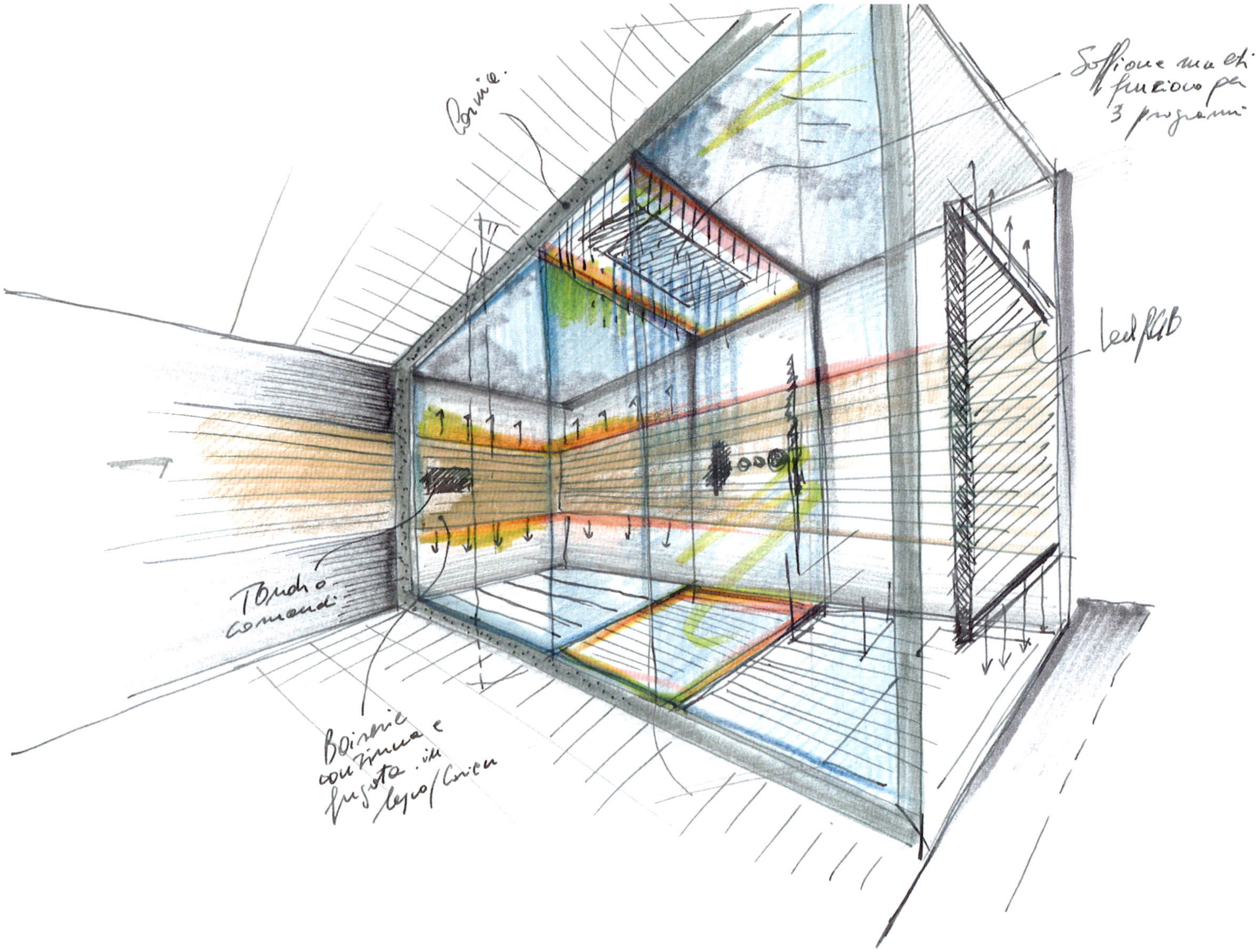
Cornice.
Soffione multi funzione per 3 programmi
Led RGB
Tondo comandi
Boiserie continua e fugata in legno/

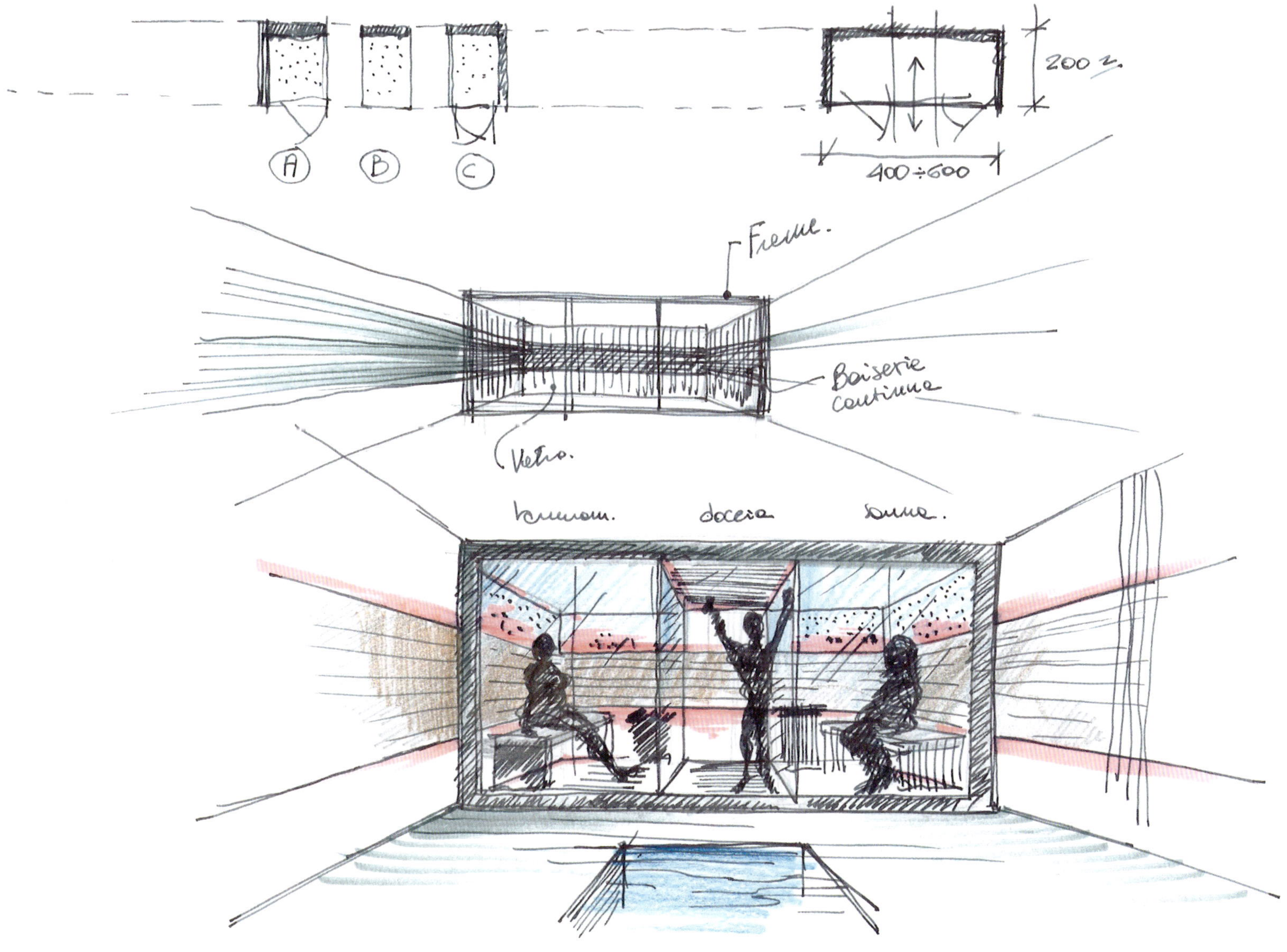
A
B
C
200
400÷600
Boiserie continua
Vetro.
hammam.
doccia
sauna.

Jacuzzi

Bliss

Title: Bliss
Design Agency: Glamorous Co.,Ltd.
Location: Kowloon, Hong Kong, China
Client: Sun Hung Kai Properties Limited
Area:489 m^2
Photography: Nacasa & Partners Inc.

introduction:

It is located on 72F and 73F of a luxurious hotel newly opened in HK. It consists of reception/retail area, manicure/pedicure room, relaxation lounges and 9 treatment rooms. Changing rooms and gym are located on 73F connected by staircase at the reception lobby.
Being inspired by the signature brand color, aqua(=water) became a design theme without any doubt. We challenged to express its ordinary theme differently in order to establish a new brand identity.

For guests to experience the brand spirits, cheerful motifs are spread throughout the space; such as acrylic drops, metal beads and original ceiling feature. Original acrylic drops glow alike pearls of dew by lighting and hydrate the space upon arrival. Metal beads at corridor give impression of gentle raindrops. Original ceiling feature at lobby is metaphor of puddle created by light. All these features combine visual pleasantness and intensions. Acrylic drops at the void connect two floors effectively as well as hiding the smoke exhaust window behind. Raindrops made of metal beads create interest for guests to walk narrow corridor without any natural light. Puddle of light at the lobby ceiling is designed to maximize its height avoiding ducts or structural beams.

Noise and privacy protections are always important for successful spa design. Original pop carpet covers the lobby and corridor to minimize the noise while offering cheerful experience to walk on. Treatment room doors are not directly opened toward the corridor. A translucent partition and glass balustrade at lobby are designed to protect the guests' privacy travelling between floors in bathrobes.

By all these efforts, we successfully created a strong brand identity while maintaining sophisticated hospitality of the hotel spa.

1

bliss: 489㎡

Reception / Retail
Manicure & Pedicure Room
Lounge
Relaxation Room
Treatment Room: 9 rooms

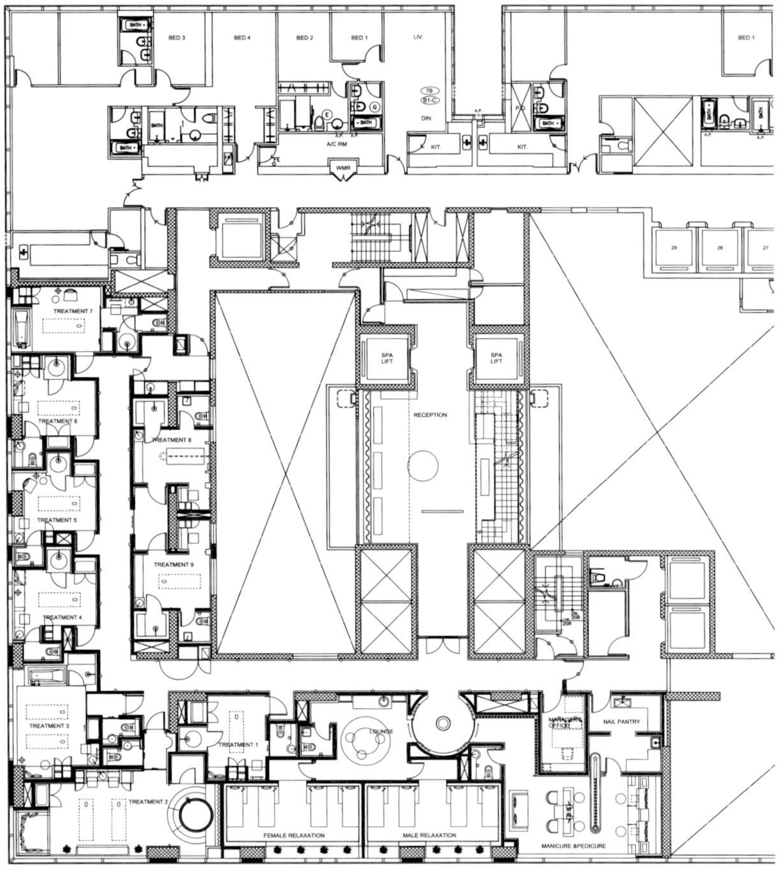

2

sweat: 136㎡
MALE CHANGING ROOM: 106㎡
FEMALE CHANGING ROOM: 114㎡

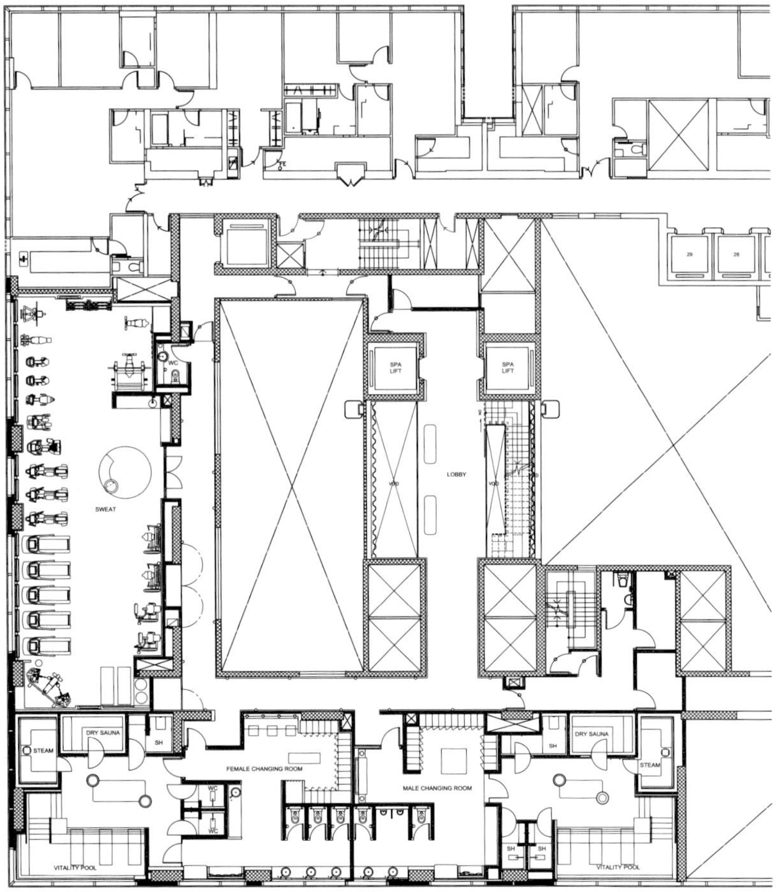

Banyan Tree Spa Al Wadi

Title: Banyan Tree Spa Al Wadi
Design Agency: Architrave Design & Planning
Location: United Arab Emirates
Client: Banyan Tree Spa
Area: 3000 m^2

introduction:

The exclusive desert resort of Banyan Tree Al Wadi located in tranquil and eco-rich Wadi Khadeja features the award-winning Banyan Tree Spa with lavish treatment pavilions and state-of-the-art hydrotherapy facilities. Set against the waters of the Arabian Gulf and the magnificent Al Hajjar Mountains, this oasis of indulgence holds true to traditional Arabic architecture, blending harmoniously with the rolling dunes of the desert. Complete with the finest décor and time-honoured spa healing therapies, Banyan Tree Spa Al Wadi is an ultimate retreat in the majestic desert. Banyan Tree Spa Al Wadi features 10 indoor treatment pavilions with twin spa beds, including six Deluxe Pavilions with attached shower room with steam and four Royal Pavilions each with a hydro-jet bathtub, outdoor shower and rain mist room for pampering tropical rain mist experiences.

EF MediSpa

Title: EF MediSpa
Design Agency: Campaign
Location: London, UK
Client: EF MediSpa
Area: 240 m^2
Photography: Hufton + Crow

introduction:

Campaign has re-designed the interior of the EF MediSpa—a small urban spa specialising in cutting-edge non-invasive surgery and beauty treatments. The unique one-stop spa on the Kings Road, Chelsea, London is the vision of beauty maven Esther Fieldgrass and offers a combination of aesthetic treatments with the latest medical advances and traditional complementary therapies.

Campaign has employed a restricted palette of materials—white Corian, glass and crystal—to convey a clinical yet luxurious feel to compliment the innovative and hi-tech treatments offered.

Spa treatments include the Drip & Chill lounge developed by Esther Fieldgrass and exclusive to MediSpa. A menu of vitamin and mineral infusions is applied through a drip to provide an instant and lasting remedy for a variety of health issues. Conditions benefiting from these infusions range from the common cold to chronic fatigue.

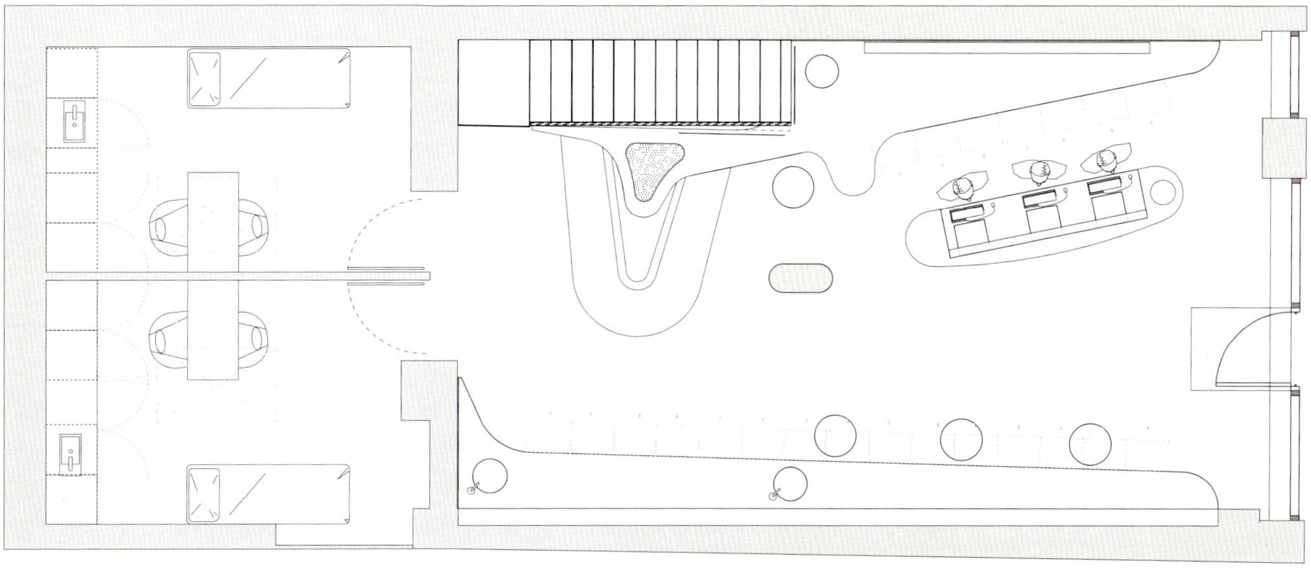

1

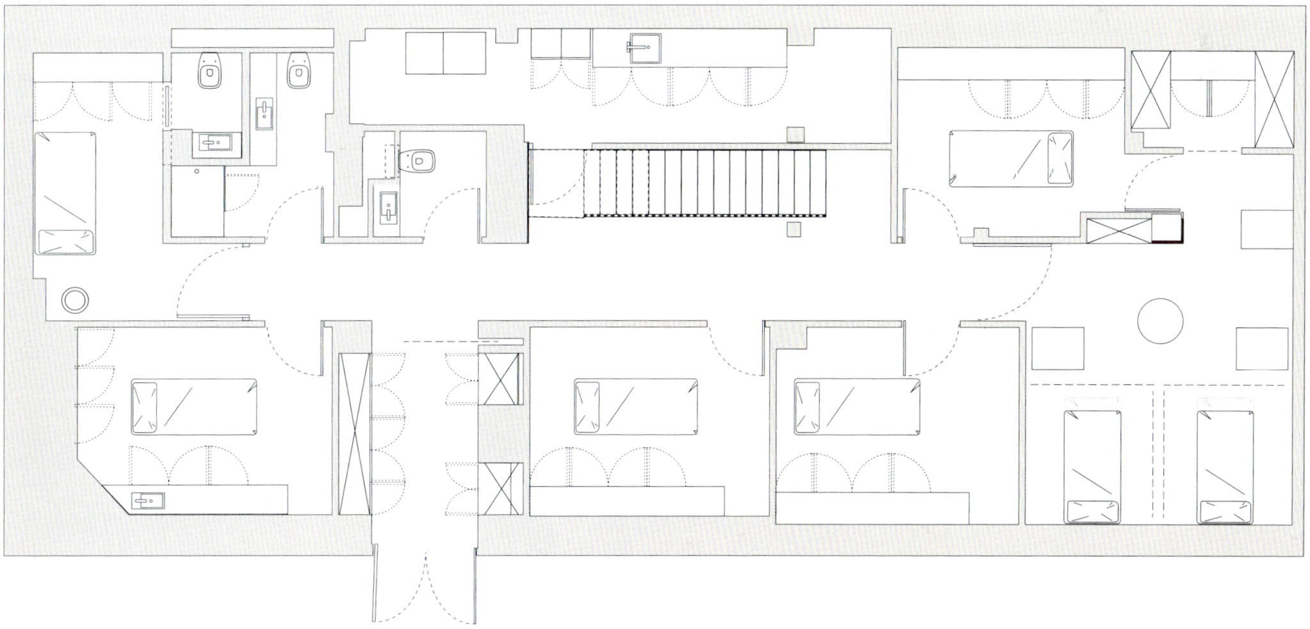

2

EFMEDISPA

skin
face
body
EF MEDISPA
INTELLIGENT SKIN CARE

EF MEDISPA

Hairu Hair Treatment

Title: Hairu Hair Treatment
Design Agency: Chrystalline Artchitect
Location: Jakarta, Indonesia
Area:70 m^2
Photography: William Sebastian

introduction:

Started with modular design and built through generous clean and simple detail, performed in natural material scheme and color, brought you to the different experience of leisure. It may look like a salon, but Hairu is a hair health care, such as treatment for losing hair, hair spa & massage.

The local point is the 2 (two) faced rough stone which also as a transition door to the wash area. The ribbon mirror on both side of the wall is carried through the view of all the space with the private sight just for the client without having eyes contact with the therapist.

The soft material, vitrage, is acting as a divider which could be moved to get along with the other. And the hard material, cotton, is as a permanent divider from the main corridor.

1

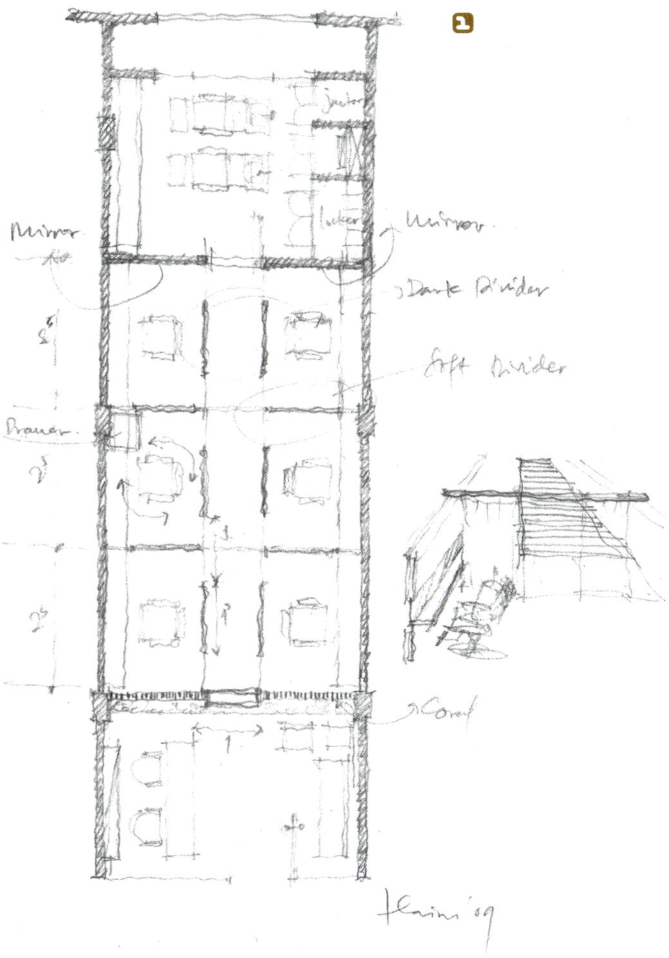

2

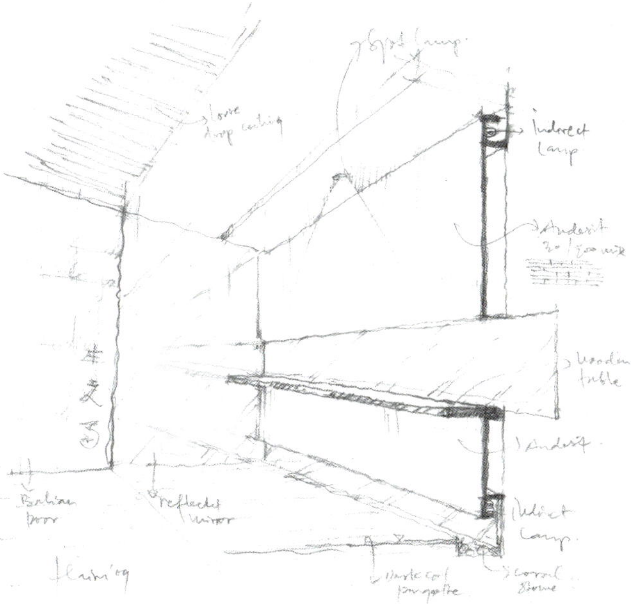

3

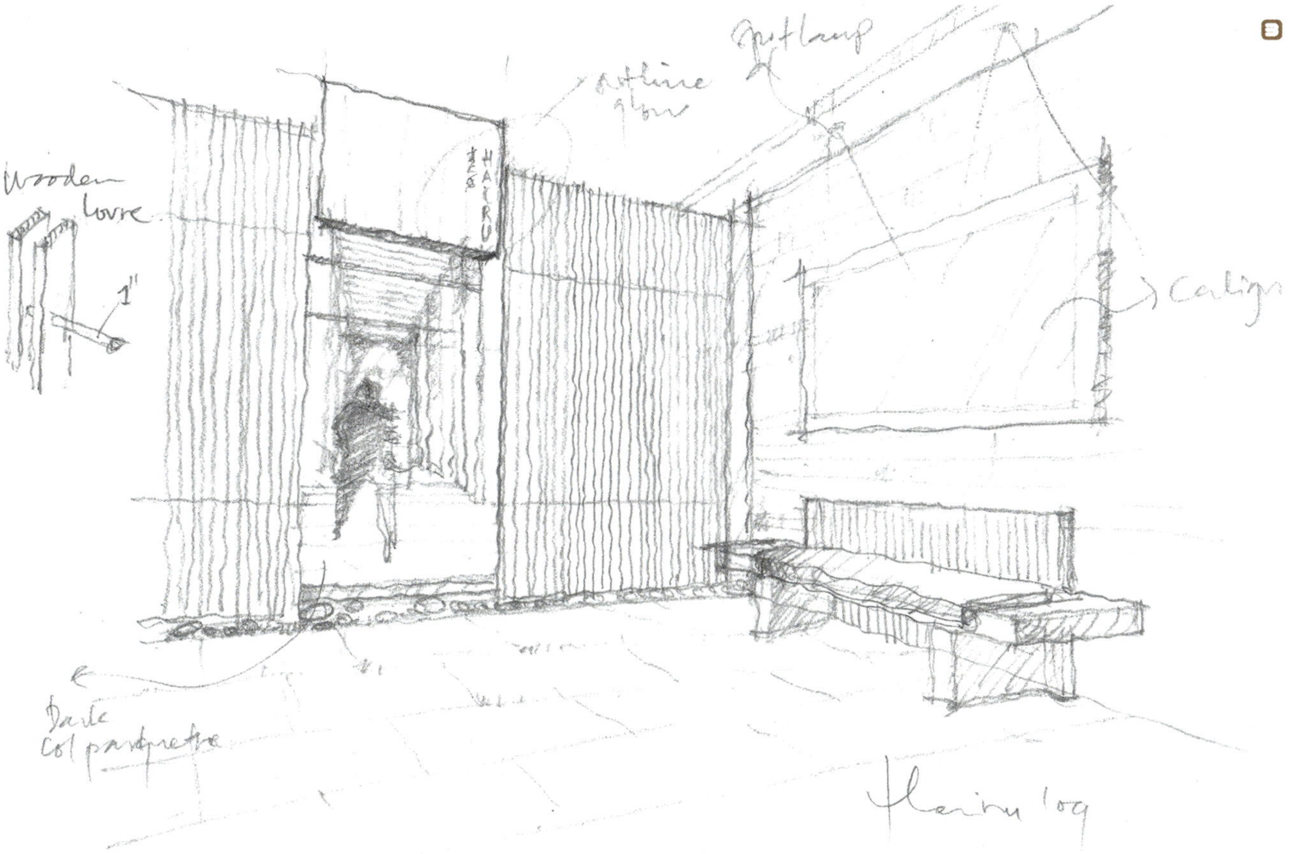

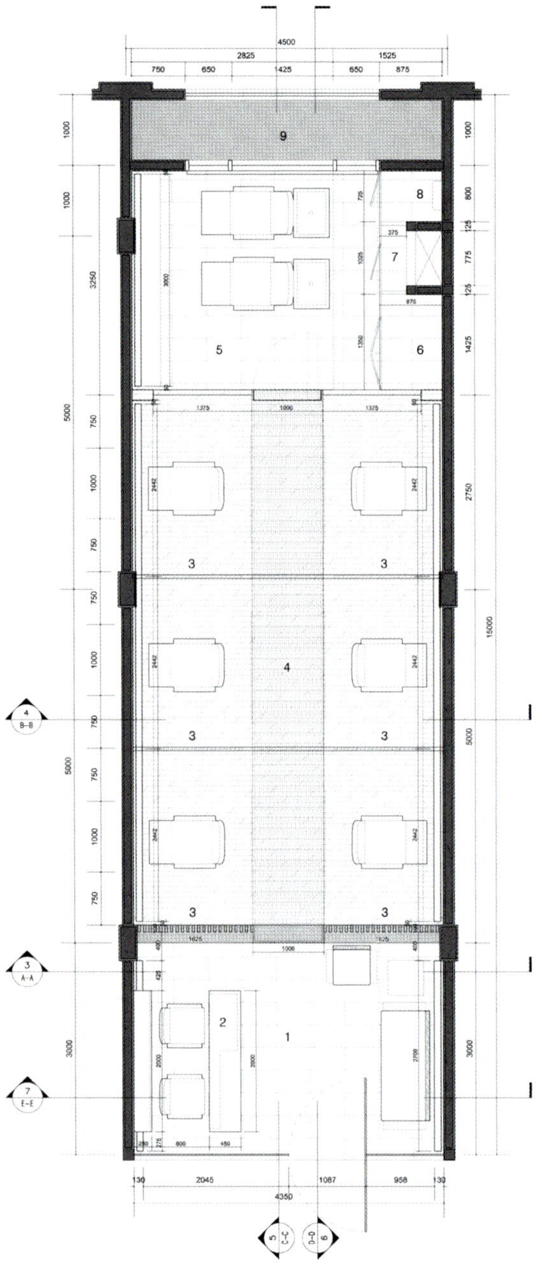

1. RECEPTION
2. CASHIER
3. HAIR TREATMENT
4. CORRIDOR
5. HAIR WASH
6. STAFF LOCKER
7. PANEL/UTILITY ROOM
8. JANITOR
9. BALCONY

1 | LAYOUT PLAN
1: 50

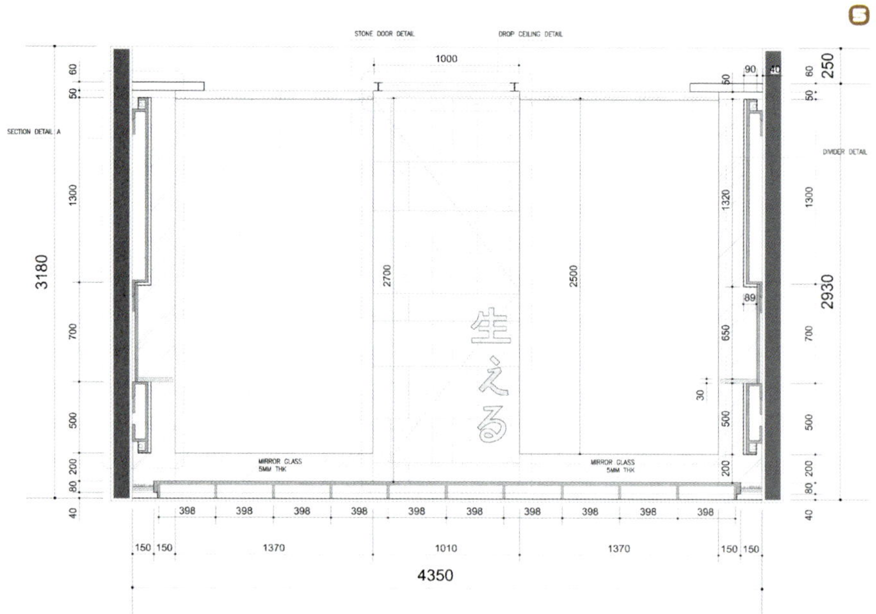

4 | SECTION B-B
1: 20

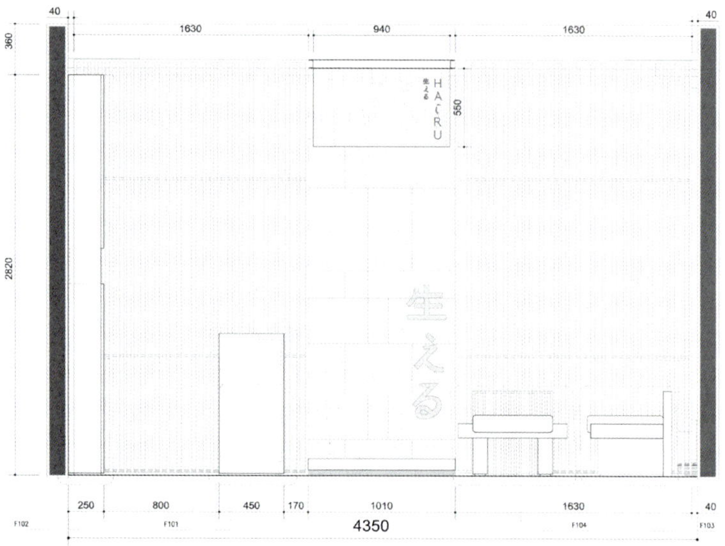

7 | SECTION E-E
1: 20

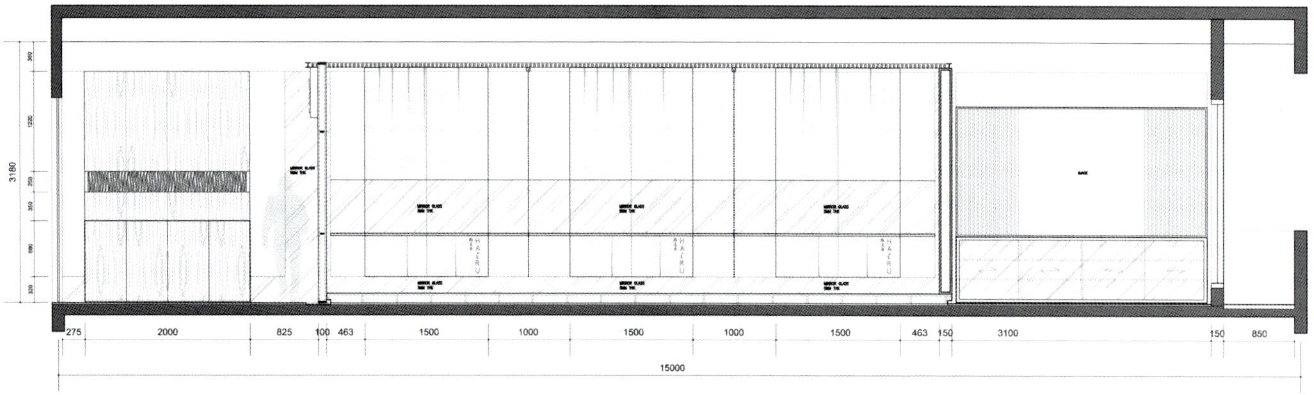

5 | SECTION C-C
1: 50

HAERU
生える

Yoga Deva

Title: Yoga Deva
Design Agency: Blank Studio
Location: Gilbert, Arizona, USA
Client: Shosh + Billy Vergara
Area: 260 m^2
Photography: Bill Timmerman, Timmerman

introduction:

Yoga Deva (deva is a Sanskrit word, meaning deity) creates an internal sequence of spaces whose primary impulse is to remove the visitor from the exterior visual environmental conditions in every way. Sited within a commercial condominium complex, the building is surrounded by asphalt parking, minimal planted islands of non-native decorative vegetation, and other buildings that are nearly indistinguishable from one to the next. This new internal environment offers an architectural and sequential chiaroscuro to the external strip-mall type reality and prepares the visitor for practice of inward meditation and contemplation.

The project exists as a series of vessels. The ancillary spaces (entry, retail, changing) are ordered and organized around the central yoga studio space. Various styles of Classical Hatha Yoga are taught at this studio, including 'hot yoga' which is a practice that is performed in a space heated to 105 degrees F to improve flexibility and detoxification benefits.

Entry occurs in a space that is visually dark in contrast to the common Arizona exterior experience and no visibility to any other space is offered. The space is long, narrow and rendered in dark pigment, dark wood walls and dark wood millwork; including a communal seating area along the length of the eastern wall illuminated by warm-toned cove lighting. Terminating the long volume is an electric blue pigmented wall washed in cool-toned light that invites you forward. The western wall of the volume's long axis is surfaced in aluminum gilding, which shimmers when lit from above, and offers soft, muted reflections of the visitors moving through the space. The space serves as orientation to the newly initiated and gathering for individuals before and after a class session.

In contrast to the entry space, the main studio space is flooded with daylight; the interplay of light is calming and ever-changing. Translucent panels diffuse the light as the memory of the arbitrary position of the glazing in the existing, exterior mall shell is muted and edited. Above, the ceiling consists of three cylindrical, inverted vaults running the length of the main axis of the space. Valleys in between the vaults collect and order the building's systems (lighting, HVAC, fire protection, etc.) and allow for light to wash across the uninterrupted vaulted surfaces.

1

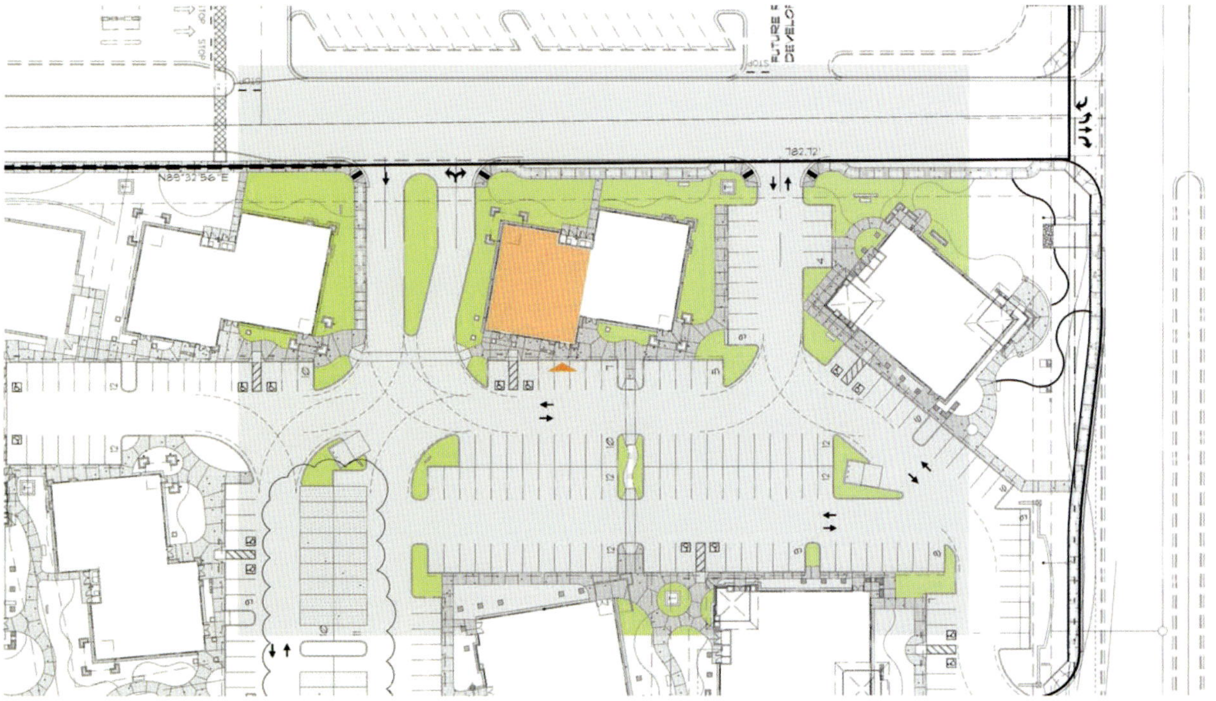

2

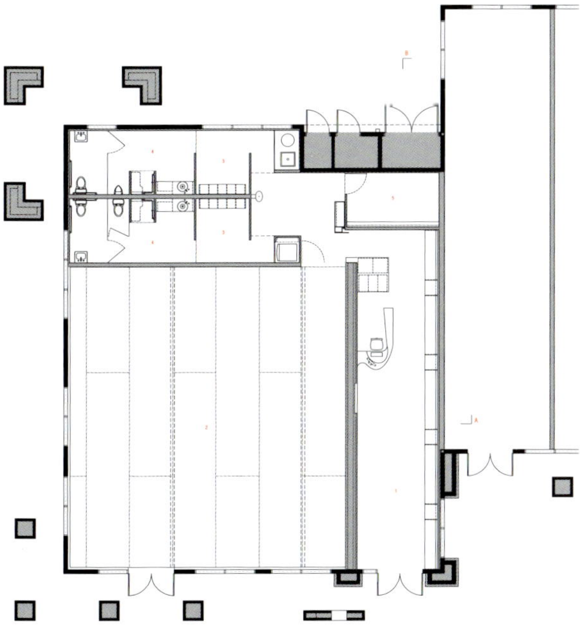

3

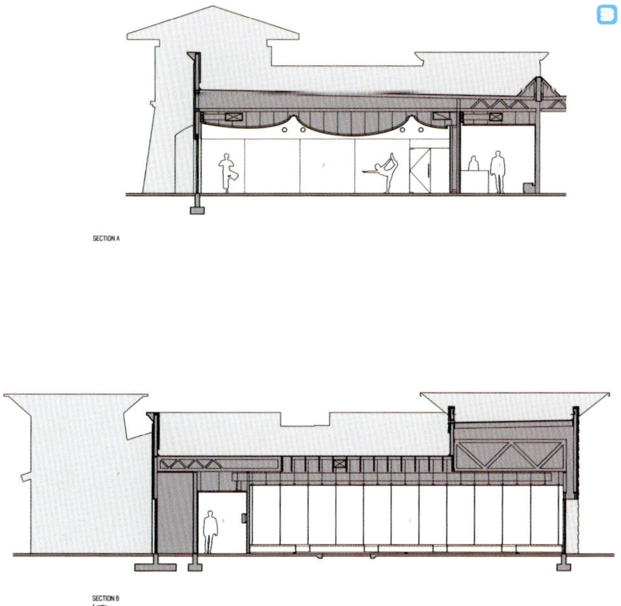

YMCA Renaissance Center

Title: YMCA Renaissance Center
Design Agency: McIntosh Poris Associates
Location: General Motors Corporate Headquarters, Renaissance Center, Detroit, MI
Client: Hines Interests
Area:1115 m^2
Photography: Michael Collyer
Mural: Barney Judge Studio

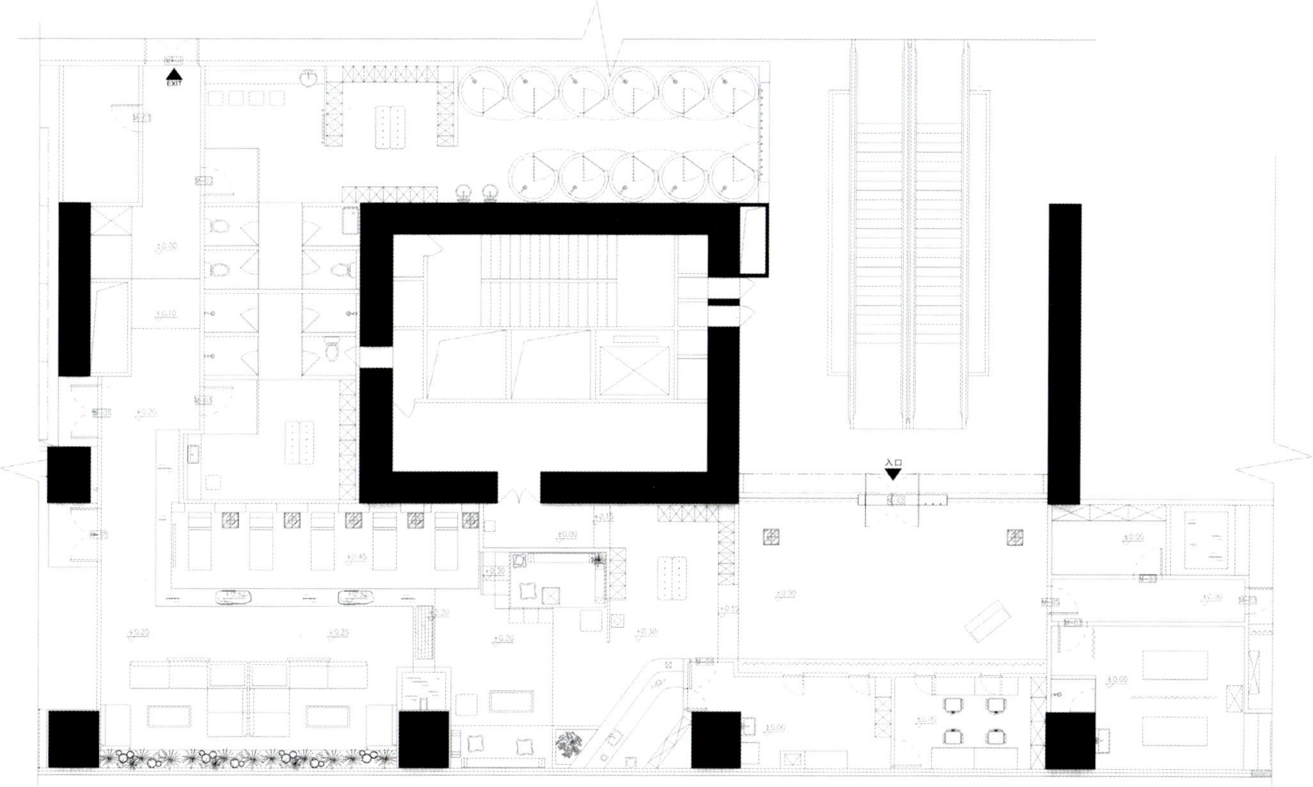

1

introduction:

Before we took over the project In Yoga Center Spa, it already had a plan and was under construction. At that moment, the proprietor showed concerns about the previous design. After communication between designer Han Song and the proprietor, he firmly believed that he had to change his plan. After modifying and adjusting, we finally had this work. It can be called a place for cheer and happiness.

This project is located in Wu Zhou Xin Yuan, Futian, Shenzhen, and occupies 1000m^2.

Due to the trust of its owner, Han Song took over it. The main element of this Yoga Center Spa design is the color gold. From the lobby and through the whole center, the designer uses triangular black and white to break the balance of the space, making it lively and active. The lighting of this design is changeable chandelier, creating abundant changes for its users. As for material, the designer uses different leather. Lighting and all the gold leather made closet doors, screens, room doors, made the Yoga center a project of exquisite charm and beauty.

SPA

消火栓

消火栓

U+ Yoga Center

Title: U+ Yoga Center
Design Agency: Shenzhen Horizon Space Design Co.,Ltd
Location:Shenzhen,China
Area:1000m^2
Photography:Han Song

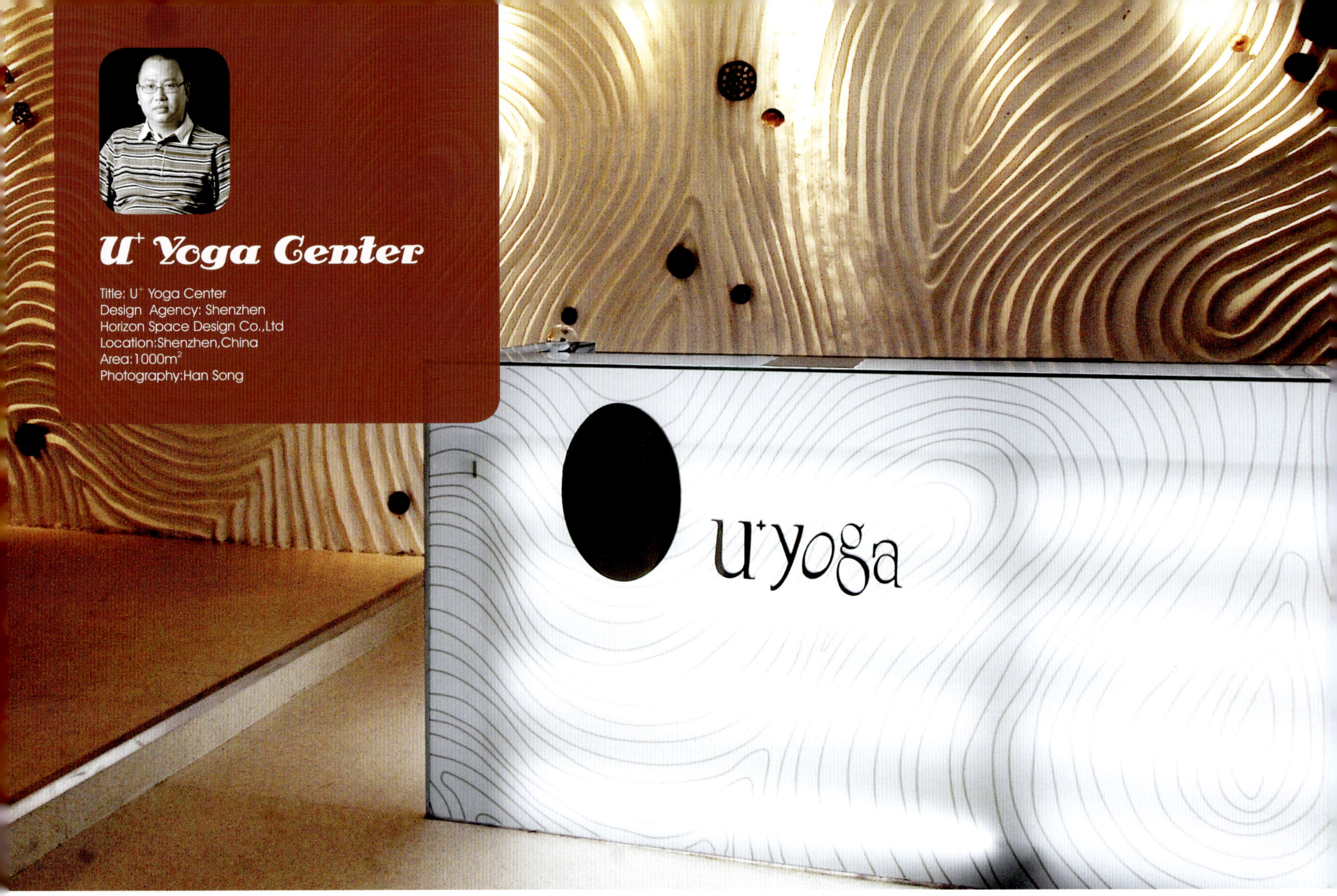

Acknowledgement

We would like to thank all the designers and companies who made significant contributions to the compilation of this book. Without them, this project would not have been possible. We would also like to thank many others whose names did not appear on the credits, but made specific input and support for the project from beginning to end.

Future Editions

If you would like to contribute to the next edition of Artpower, please email us your details to: info@artpower.com.cn.

全球顶级水疗设计

THE WORLD SPA DESIGN

养生中心 美容中心 酒店水疗 室内设计 1/2

HOTEL SPAS & BEAUTY SPAS \ WELLNESS CENTERS INTERIOR DESIGN

Vol.2

大连理工大学出版社

深圳市艺力文化发展有限公司 编

图书在版编目(CIP)数据

全球顶级水疗设计. 1、2 : 英、汉 / 深圳市艺力文化发展有限公司编. — 大连 : 大连理工大学出版社, 2011.8

ISBN 978-7-5611-6325-2

Ⅰ. ①全… Ⅱ. ①深… Ⅲ. ①保健建筑—建筑设计—作品集—世界 Ⅳ. ①TU246.9

中国版本图书馆CIP数据核字（2011）第134380号

出版发行：大连理工大学出版社
（地址：大连市软件园路80号　邮编：116023）
印　　刷：利丰雅高印刷（深圳）有限公司
幅面尺寸：245mm × 290mm
印　　张：32
插　　页：8
出版时间：2011年8月第 1 版
印刷时间：2011年8月第 1 次印刷
责任编辑：初　蕾
特约编辑：李　莹
责任校对：仲　仁
装帧设计：杨思雁

ISBN 978-7-5611-6325-2
定　　价：560.00元（共两册）

电 话：0411-84708842
传 真：0411-84701466
邮 购：0411-84703636
E-mail：designbook@yahoo.cn
URL：http://www.dutp.cn

如有质量问题请联系出版中心：（0411）84709246　84709043

PREFACE

1

What is it that drives us to design spas?

In the simplest of terms, it is the desire to dream and to enhance the wellness of others.

It starts by 'sinking-in' or grounding ourselves within the project. This involves understanding the locale, history, customs, and healing practices that are native or unique to the area and the market for which the facility is built. The mere process of gaining this type of knowledge leads to a mental, physical and spiritual journey that encourages dreaming.

Dreams are organic and multi-dimensional, involving the senses, emotions and feelings. The approach is to imagine the experiences that the individual will ultimately have while enjoying the spa. What does the facility look like? what is the experience? What is the mood, the feeling, the character? These mental pictures create the reference points that shape the facility program and design. Color, form, function, and light emerge. Ideas swirl, images appear and plans are created.

A tapestry is made of the architectural and interior design, special touches, spa menu, wellness program, creating the overall spa journey. Every component, from facility to experience, has a unique element, and functions in complete harmony.

Finally, the spa comes alive with the arrival of staff and the spa guests. The dream of life enhancement becomes a reality that focuses on wellness, with knowledge and tools that create lasting experiences long after their journey at the spa. And the measure of a successful design is when owners and guests alike find the spa a profitable investment of their precious time and money.

Whether grand or simple, planning and designing spas is a dream come true. In this book we cherish the privilege to share the pages with and to witness the dreams of other designers.

Paige Megna, Principle
SPAd Sylvia Planning And design

2

People have different opinions regarding the term 'SPA'. Some affirm it derives from the Latin expression sano per acqua, which means 'healing with water', whereas some affirm its origin is the name of a Belgian town, named 'Spa'. Regardless of its etymology, most people agree that the word sends us back to the thermal baths of the classic old times, when Romans first experimented those baths, not only as relaxation rituals, but also for socializing purposes.

Therefore, Spas have always had a direct connection to water-related therapies in the first place, i.e. thermal therapies in hydro-mineral venues, offering fruition of waters with medical properties. This is the origin of nowadays hydro-therapeutic treatments using water, steam or infusion, which have been complemented with massage and other non-intrusive medical treatments throughout the years.

Presently, day-by-day life turbulence is drawing out our vital energy. Extreme work routines, the speed with which information comes and goes and responsibilities building up are some of the aspects which could potentially dry our body and mind up. Due to such wearing, we are prone to seeking some form of compensation, as bodies and minds urge for relaxation. That is indispensable.

The physical, mental and emotional benefits of treatments and therapies as they are offered by the Spas, in addition to one's welfare, could alleviate accumulated stress, calm brain and nerves, stimulate the immunologic system, soothe aging, maximize self-esteem, concentration and creativity, cleanse the body of toxins, prevent and treat muscle problems, among many other benefits.

Maintaining a sound body and a sound mind has always been the major goal of any Spa. They are most of all, a refuge from stress and daily fatigue. In that sense, Spas privilege treatments dedicated to physical, psychic and aesthetic welfare, aimed at men and women of all ages. They become hence a significant part of the lives of those who attend it, once positive effects are observed in life quality, self-esteem and harmony between body and mind. That's why it is so important to know what is of major in terms of spa, so we can apply it in to a space and project a dreaming spa, capable to offer all the relaxation that we all need.

With technology enhancement, contemporary Spas were able to equip themselves with different therapy techniques, varying from alternative/holistic treatments, such as acupuncture, to various body and facial treatments, e.g. skin cleansing and exfoliation, face masks, body wrap in honey, milk, chocolate or wine, relaxing massages etc.

Today's Spas display ultra-modern or pretty traditional facilities and equipment, including hot tubs, swimming pools, saunas, Turkish Baths, relaxation chambers, treatment chambers, aesthetics chambers, beauty salons, between others. SPAs deco and visual aesthetics play a fundamental role in bringing a comfortable and most importantly relaxing environment. All details are pretty fundamental. Nothing is left to chance. Colors, textures, textiles such as towels and bathrobes, slippers, candles, flowers, plants, decorative items and background music, it all comes down to the physical and mental welfare of those who will enjoy the pleasures offered by a Spa.

In this book you will get the chance to see dazzling landscapes from the global and local city picturesque sites, introducing a vast number of the most beautiful and relaxing Spas in the world. This book displays the hottest picks when it comes to state-of-the-art innovating resort spas, hotel spas, beauty spas, yoga clubs, hair spas and a lot more. Do not pass this chance up, make your choice and relax!

Mariana Falcão & Daniel Corrêa
auCUBO.design | architecture + urbanism

016
Akatsuki Resort
208
Spa Village
Pangkor Laut
Mii amo Spa
044
Thermal Römerbad
Spa Bath Bad Aibling
102
Tschuggen
Berg Oase

DIRECTORY

024
Capella
Singapore
178
Eco Spa

Titanic
Mill Spa
134
Cordillera
170
084
Hotel Post
224
Angsana Spa Hotel
Nikko Shanghai
Asian
Roots Spa

246
Shenzhen Jingyuan Yoga Studio
008
Wellness Orhidelia
158
Joya Spa
094
The Colony
252
Shenzhen In Yoga GYM
202
Angsana Spa Tivoli Marina Vilamoura
110
Spa 'Casa Velha do Palheiro'
118
Bella Collina
182
Spa Mais Vida
060
Emirates Resort & Spa
Xi Er Si Chi Dian
150
Pronghorn
142
Pine Canyon
126
The Club at Black Rock
234
Dolder Grand Spa

Wellness Orhidelia

Title: Wellness Orhidelia
Design Agency: Enota
Location: Podčetrtek, Slovenia
Client: Terme Olimia
Area: 9990 m^2
Photography: Miran Kambič

Akatsuki Resort

Title: Akatsuki Resort
Design Agency: Riccardo Tossani Architecture
Location: Lipa Noi Beach, Koh Samui, Thailand
Client: Chartered Holidays Ltd.
Area:950 m^2
Photography: Martyn Goodacre

Capella Singapore

Title: Capella Singapore
Architect: Foster + Partners
Location: Sentosa Island, Singapore
Photography: Robert Reck

introduction:

Capella Singapore, the flagship property for Capella Hotels and Resorts in Asia offers an ultra-luxurious personal sanctuary just minutes from Singapore's financial and shopping districts. Capella Singapore promises the ultimate in personalized service and represents a new standard of luxury in Asia, combining the best of old and new Singapore. From arrival to departure, Capella Singapore cossets its guests, delivering intuitive and uncompromising service. Personal Assistants expertly handle guests' needs and private chauffeur services are available via the resort's Rolls Royce Phantom.

Capella Singapore is a proud edition to the elite design profile of famed architects Foster + Partners, London. Thoughtfully designed, the resort is shaped like a figure-eight, an auspicious number in the Chinese culture that signals prosperity. The property slowly unveils itself, revealing at first only a hint of the curving architecture that cascades down towards the sea.

Guests can also enjoy a range of activities for relaxation or business. Cassia, the fine-dining Chinese restaurant, The Knolls, the all-day dining restaurant and Bob's Bar. Auriga Spa and a Fitness Center are also available, in addition to three meeting rooms; a business center with two boardrooms, a Library and three cascading outdoor pools.

Capella Singapore also offers guests the opportunity to live on the property for extended periods with full access to the resort's facilities. These extended stays include seventy-two sea-facing suites and duplexes and nine manors with private pools.

Mii amo Spa

Title: Mii amo Spa
Spa Design: SPAd (Sylvia Planning And design)
Architect: Gluckman Mayner Architechts
Interior Design :Glauckman Mayner Architechts
Location: Sedona, Arizona
Client: Enchantment Group
Area: 3158 m^2

Introduction:

Located in Sylvia's home town of Sedona, Arizona, Mii amo is a destination spa at Enchantment Resort. Sylvia envisioned this stunning facility as a retreat for guests on a journey of transformation and exploration. Inspired by the rich wisdom and traditions of Native American people, complemented by the beauty and energy of Sedona, this destination spa resort presents a luxurious spa experience where guests can find balance and harmony in breathtaking Boynton Canyon.

The creation of this unique Arizona destination spa was inspired by one of the most spectacular natural settings on earth, Boynton Canyon, and by the Native American culture that graces the region. The result is a fusion of Arizona-inspired colors and design elements, contemporary architecture in a pristine environment. The two-level main building is nestled into the natural slope of a red-rock canyon wall. Simple lines, monumental forms and pure materials such as adobe brick, wood details and indigenous stone blend organically with the surroundings of this Arizona destination spa.

Natural light and water are recurring themes, and interior and exterior spaces flow into one another. Communal areas and pools are on the ground level, with treatment rooms housed in five adobe brick towers. Native American-inspired colors, design elements and hand-woven rugs, with bamboo and terrazzo floors, enhance the architecture of Mii amo.

Some of the unique facilities at Mii amo include the Crystal Grotto, indoor and outdoor pools, the demonstration kitchen, outdoor treatment Wikiups, and indoor treatment rooms with angled windows for red rock canyon viewing.

Spa Entrance →
← Casitas 501-516

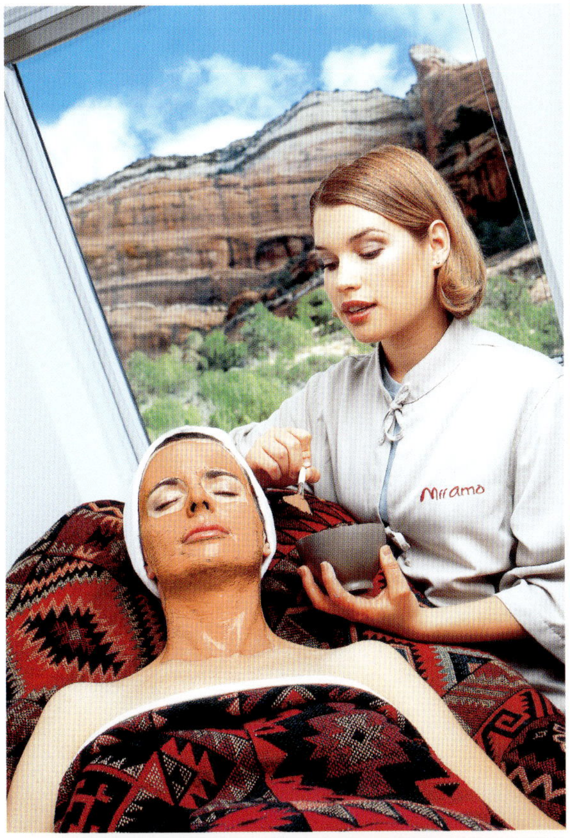

Thermal Römerbad

Title: Thermal Römerbad
Design Agency: Behnisch Architekten—Stefan Behnisch,David Cook,Martin Haas
Location: Bad Kleinkirchheim, Austria
Client: Bad Kleinkirchheimer Thermen GmbH
Photography: Roland Halbe

Spa Bath Bad Aibling

Title: Spa Bath Bad Aibling
Design Agency: Behnisch Architekten—Stefan Behnisch,David Cook,Martin Haas
Location: Bad Aibling, Germany
Client: Stadtwerke Bad Aibling
Photography: Adam Mørk

3

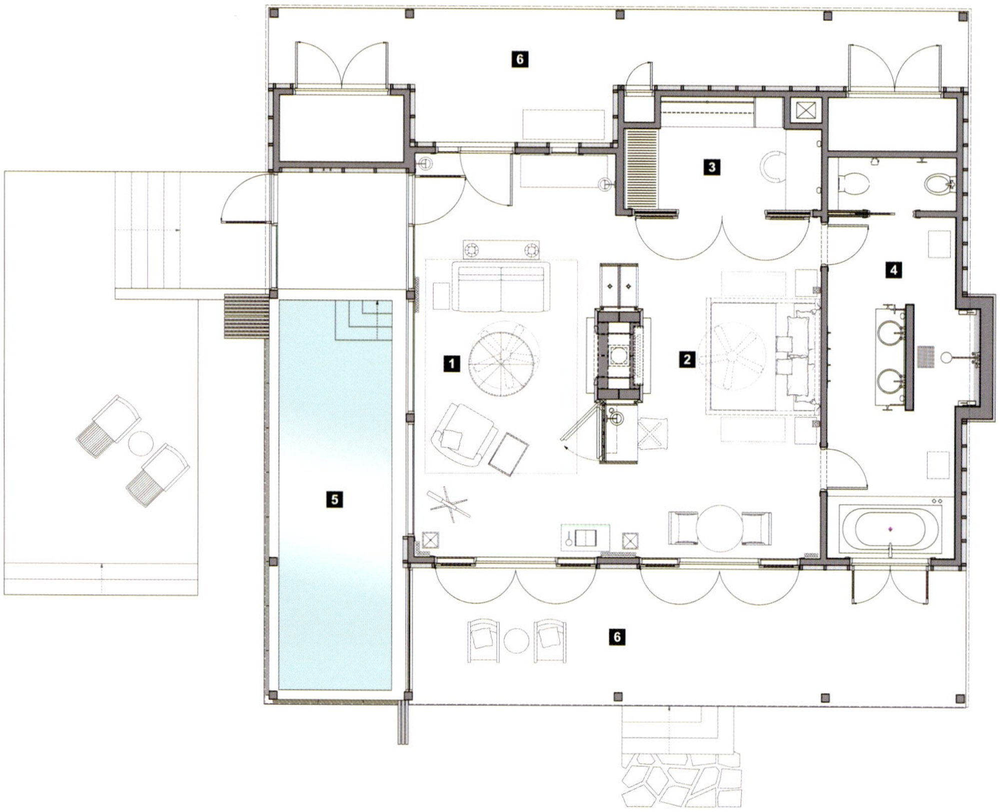

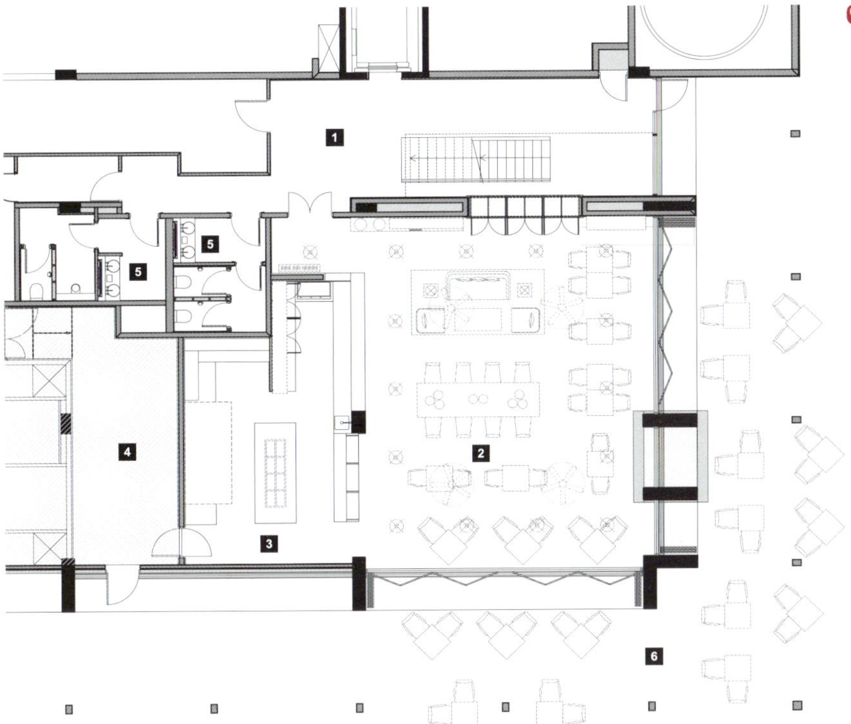

4

Soupe du Jour
Potato and leek

morphogenesis.

Asian Roots Spa

Title: Asian Roots Spa
Design Agency: Morphogenesis
Location: New Delhi
Client: M/S Asian Roots
Area: 595 m²
Photography: Amit Mehra & Andre J Fanthome

introduction:

Taking its inspiration from the birth of holistic treatments in Asian traditions the Asian Roots Spa, is designed as a tranquil sanctuary. The spaces designed in a minimalist palette provide a serene sense of calm and well being. The entry to the spa is through a zen garden with a Japanese dry landscape garden or Karesansui forming the focus of the entry. A bottle palm forms an asymmetrical natural accent to the space. Wooden seats placed on a floating stone bench provide a place for contemplation in this soothing setting.

Entering through the wooden floored reception, one comes to the heart of the spa, a large triple height atrium space. Light floods into the atrium diffused through billowing fabric through a skylight three floors above. The space overseen by a large wall painting of the Buddha links the spa at the various levels, through internal windows, and courts. The treatment rooms on the ground floor overlook the zen garden providing a connection with nature even in this small urban site. The interior details extend the space of the garden inwards. Free-floating circular mirrors de-construct the walls, reflecting fragments of the inner spaces. A glass floor recreates the dry landscape zen garden within the spa.

The treatment rooms on the upper floors are reached via a dramatic cantilevered staircase rising through the atrium. The treatment rooms are highly individualistic and minimalistic while retaining the functional aspects of them, Balinese and Malaysian treatments. Creation of a calm and soothing ambience is extended through a restrained palette of materials chosen for their sensory and tactile qualities, with texture and color used to provide luxurious accents. Lighting controls synchronized with treatments heighten the experience. The yoga room is a composition of all white surfaces with a light wood floor. Translucent skylights bathe the space in an even natural white light.

1

1. FRONT COURT
2. RECEPTION
3. CIRCULATION
4. RECEPTION-2
5. COURTYARD
6. BEAUTY SALON
7. MAKEUP SPACE
8. WAXING ROOM
9. STORE
10. PANTRY
11. STAIRA
12. RARE COURT
13. SERVICES

1. CIRCULATION
2. MASSAGE / FACIAL ROOM
3. YOGA ROOM
4. TOILET
5. CHANGING ROOM
6. SHOWER
7. OFFICE
8. STAFF ROOM

1. CIRCULATION
2. MASSAGE/ FACIAL ROOM
3. RANI SUITE
4. TOILET
5. CHNAGING ROOM
6. SHOWER
7. STEAM ROOM
8. STAFF ROOM
9. PANTRY/ STAFF AREA
10. TERRACE

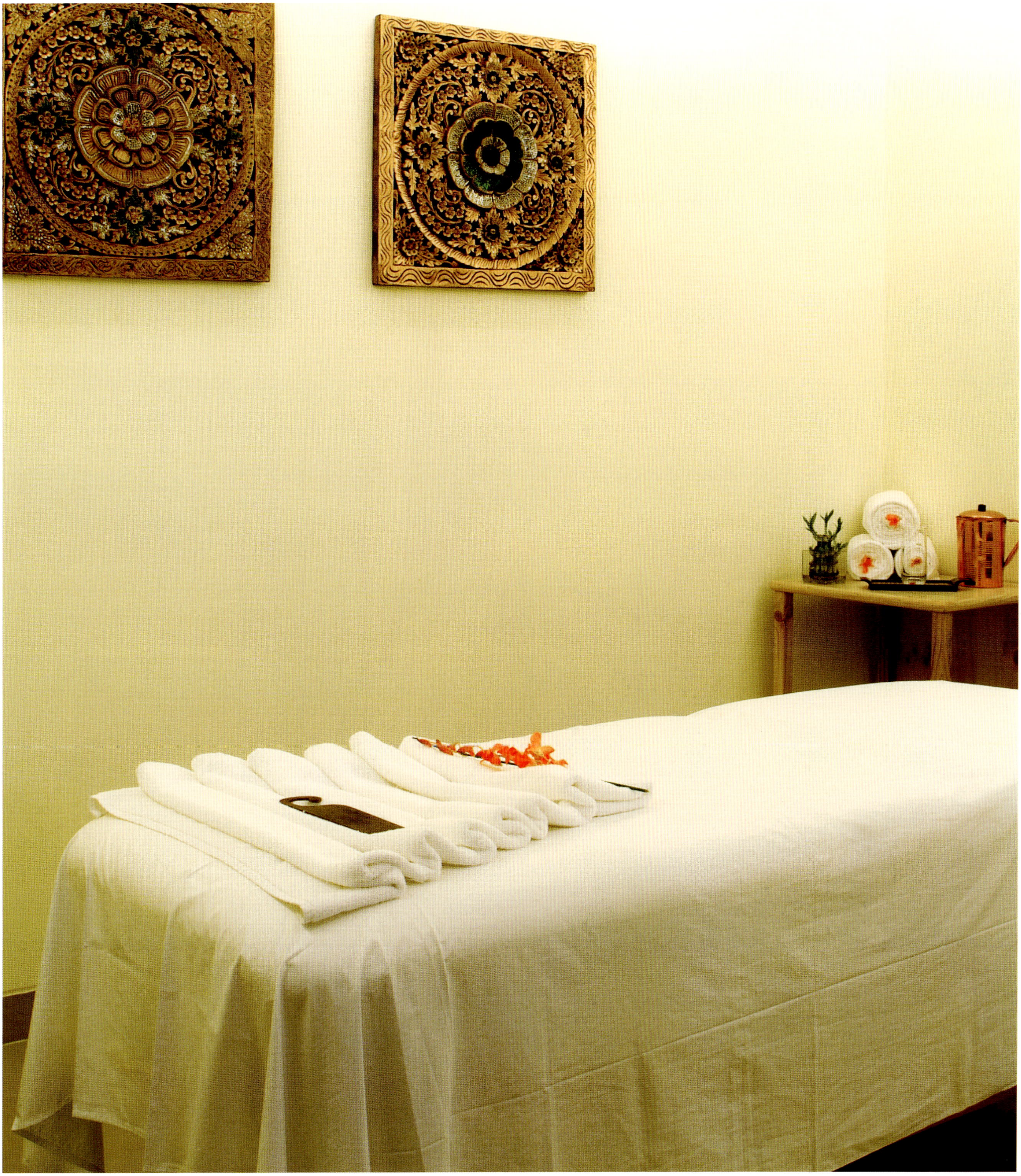

Hotel Post

Title: Hotel Post
Design Agency: Oskar Leo Kaufmann & Albert Rüf
Location: Bezau, Austria
Client: Hotel Post
Photography: Alexander Gnädinger

introduction:

We honor the old, welcome the new and remain true to ourselves and our country. That is the motto that exactly fits the spa at the Hotel Post in Bezau. Craftsmen from the Bregenzerwald, modern architecture and local materials define the design. Together with her brother Oskar Leo Kaufmann, whose design for a prefabricated building in 2008 at the New York MoMa caused a stir, Susanne Kaufmann developed the distinctive architecture of the hotel. The Hotel Post in Bezau stands for sustainable luxury tourism, the highest standards of environmental protection and meets the unique landscape of the Bregenzerwald reflects and embodies the culture of the region.

The guests of the Hotel Post appreciate luxury and want neither to give more authenticity to sustainability. The architecture of the hotel fits perfectly into the landscape of the Bregenzerwald, which is no surprise, as the architect Oskar Leo Kaufmann comes from Bezau and is with his business partner Albert Rüf, one of the main protagonists of the Modern Architecture in Vorarlberg.

The bath house with a Finnish sauna, bio sauna, steam room, sauna and indoor garden invite you to take a break. The garden and sun deck provide a view of a pristine, stunning mountain scenery. Visionary thinking and nature come together in the Susanne Kaufmann spa at a resort for all the senses.

Luxury as a focus on the essentials. The purist, all in white held spa makes it easy to let go and relax. With a lot of white and clear, almost delicate forms are the light-filled rooms of the spa is the ideal framework for face and body rituals after Susanne Kaufmann™ organic treats. We do not see luxury and naturalness as a contradiction, thinking innovative and re-define the word wellness. Relaxation, deceleration and harmonization of body, mind and soul.

1

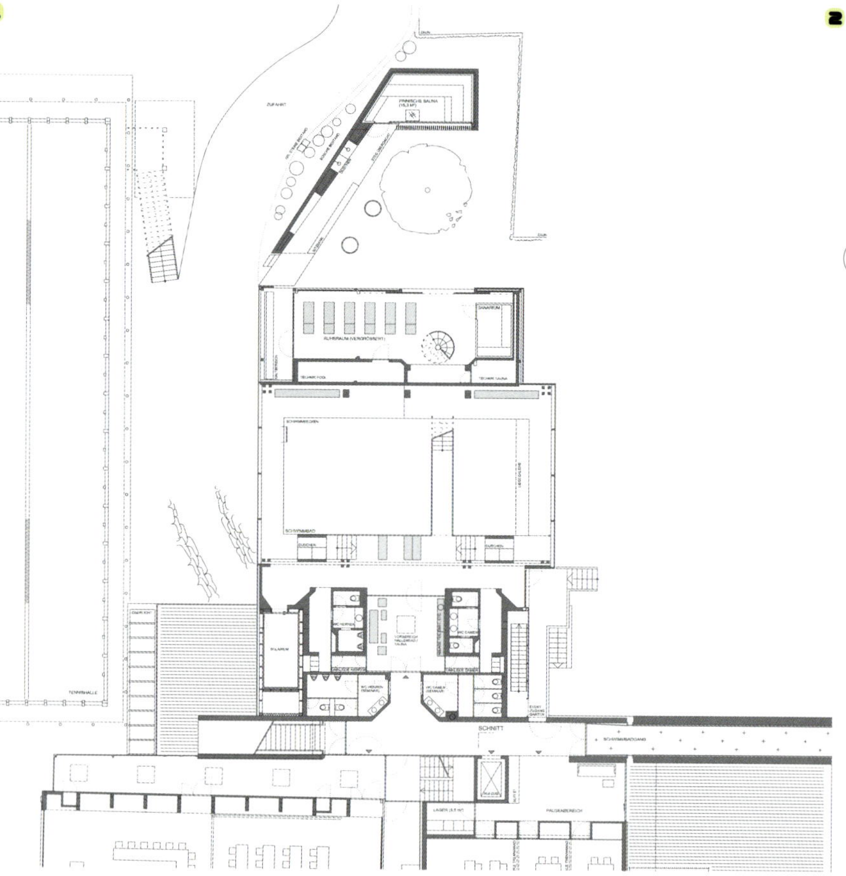

Grundriss E0 1:200

2

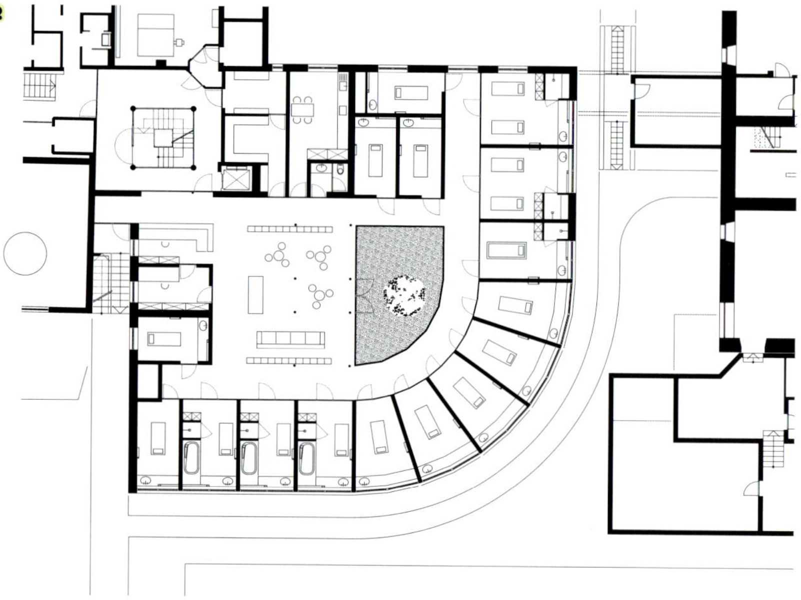

3

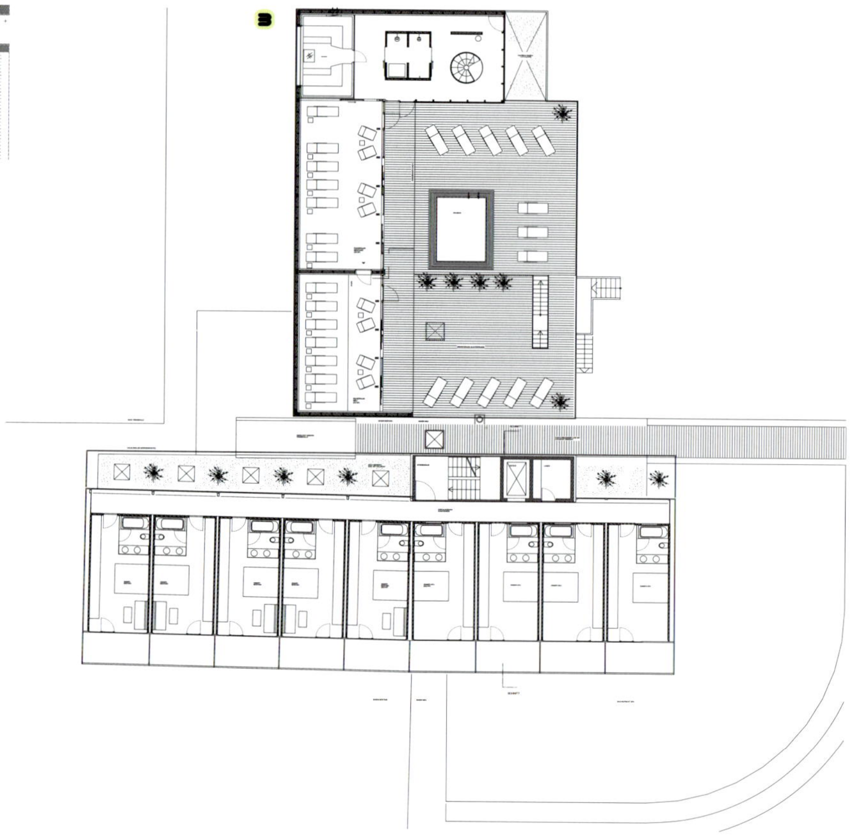

4

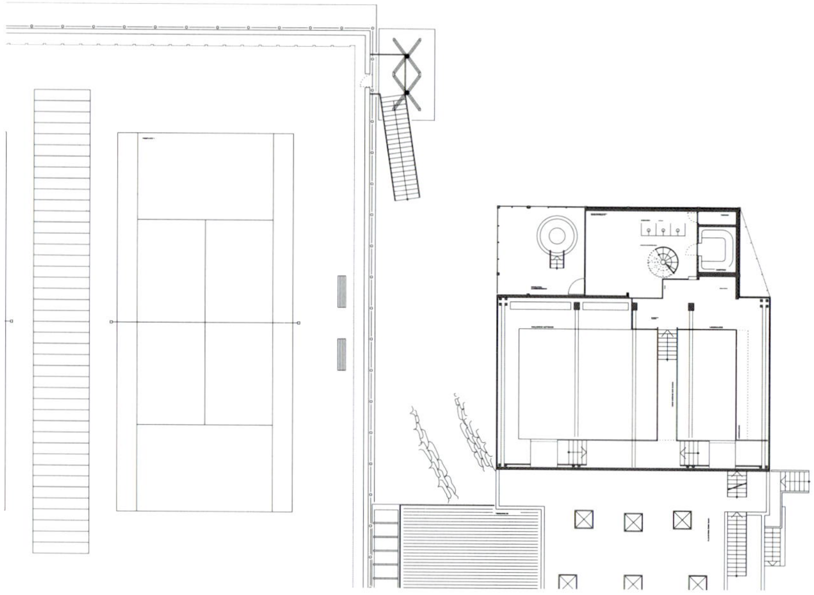

SUSANNE KAUFMANN
Anti Aging
SUSANNE KAUFMANN
Anti Aging

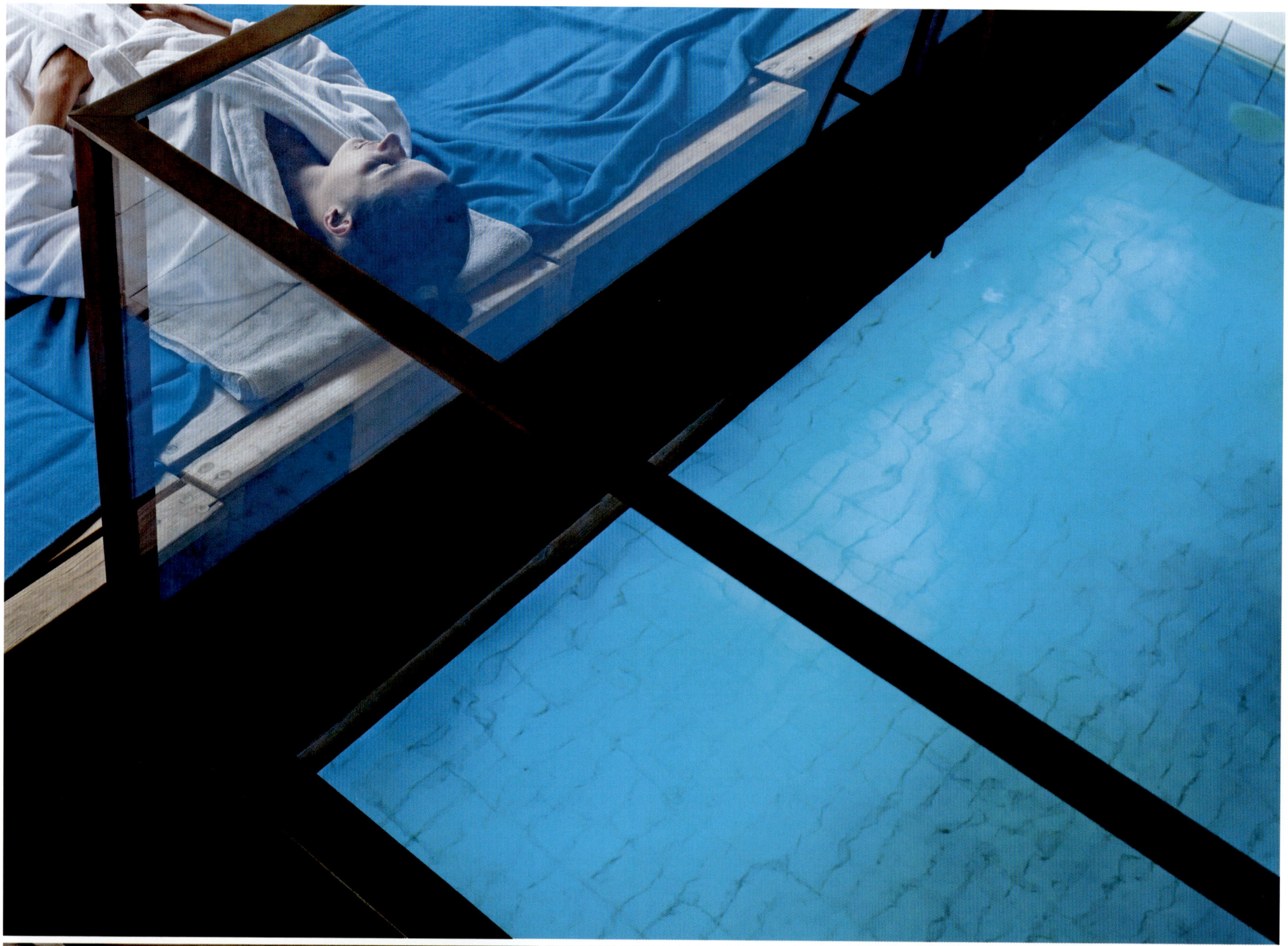

The Colony

Title: The Colony
Design Agency: ICRAVE
Location: Hollywood, CA
Client: SBE Entertainment
Area: 12,800 m^2
Photography: John Ellis

introduction:

As if a delectable slice of the Hamptons were cut out and served to the incandescent Hollywood scene – The Colony provides a sense of traditional, east coast holidaying that the west coast has been lacking. This 13,000 square feet venue located on Cahuenga Boulevard is the former Halo and White Lotus and is the latest and greatest of SBE Entertainment's night life destinations.

The design concept drew from the Hamptons beach resorts and their famous day-into-night summer house parties. The Colony is designed as a continuous wooden deck that moves seamlessly inside and outside the building. A multi-level circuit is created around the club so that every vantage point is unique and keeps the crowd wanting to explore beyond the dance floor or their private table. Blurring the edges between inside and outside was also key, so the building wraps around a large outdoor patio with a water feature at its center. 10 feet tall wooden louvered doors slide open at the dance floor edge and disappear completely, extending it into the patio.

The material palette defies the typically slick finishes of most nightclubs. Distressed barn wood was salvaged to clad the exterior and interior floors, walls and seating. Bright and striped outdoor fabric pillows are stacked on two-tiered platform seats that surround the DJ booth and dance area. In the main bar space, thick layers of brightly colored paint coat the seating areas known as 'the cabanas'. The furniture, designed by ICRAVE combines weathered materials like wood and aged steel and no distinction was made between what goes inside and outside, it is all conceived as one environment. From the street, the buildings' weathered wood walls bury themselves into large sand dunes planted with soft beach grasses. A tall structure reminiscent of a lifeguard station signals the main entry into the club through its patio.

Another element intended to release inhibitions is the Market, a retail concept that will sell props and accessories to guests that want to assume a new attitude for the night: sunglasses, whistles, scarves, face paint, flip flops, sun tan oil… The perfect summer accessories. The Market will also feature revolving follies such as a henna tattoo artist, a fortune teller and a masseuse.

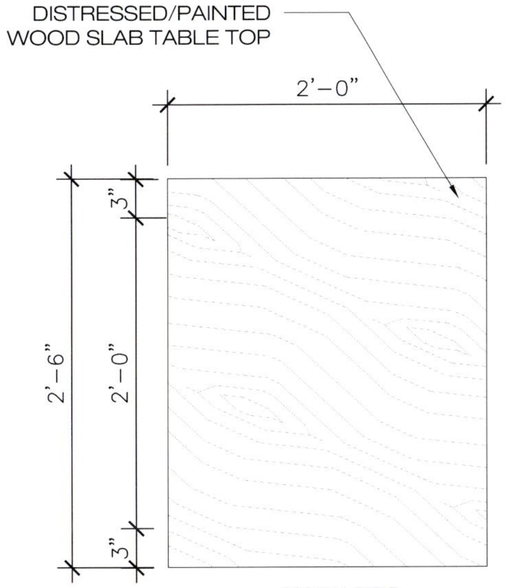

4 COCKTAIL TABLE
PLAN
SCALE: 1"=1'-0"

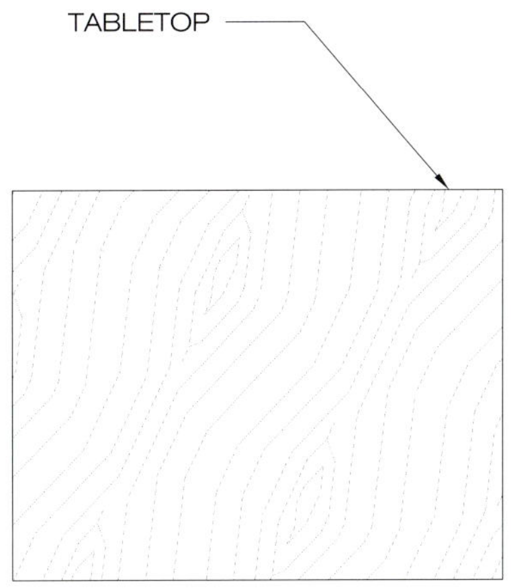

3 COCKTAIL TABLE
PLAN
SCALE: 1"=1'-0"

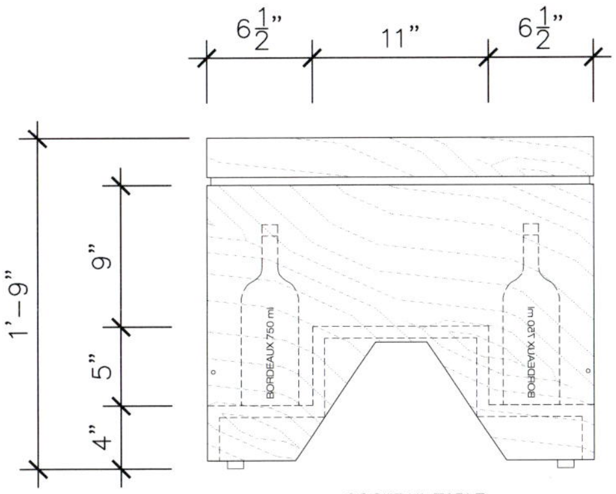

2 COCKTAIL TABLE
ELEVATION
SCALE: 1"=1'-0"

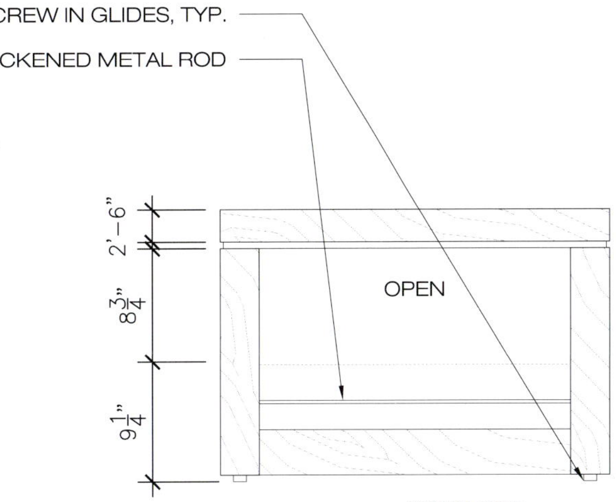

1 COCKTAIL TABLE
ELEVATION
SCALE: 1"=1'-0"

COLONY

185

Tschuggen Berg Oase

Title: Tschuggen Berg Oase
Design Agency: Mario Botta Architetto, Lugano, CH Partner architect
Location: Arosa, Switzerland
Client: Tschuggen Grand Hotel, AG Grandhotel Tschuggen
Area:5300 m^2
Photography: Urs Homberger, Enrico Cano

introduction:

Arosa offers an extraordinary geographic configuration of natural basin surrounded by mountains.

The great volume disappears into the earth and only the skylights ('trees of light') emerge. The different areas of the center are characterized by the modular design that allows their interrelation and privileged relationship with the environment.

The interior space is divided into four levels and the external sauna, solarium and swimming pool are reachable directly from the swimming pools and set on an attractive terracing, dipped into nature.

The new structure is accessible through a glass walkway from the existing hotel as well as from the entrance level to the hotel.

1

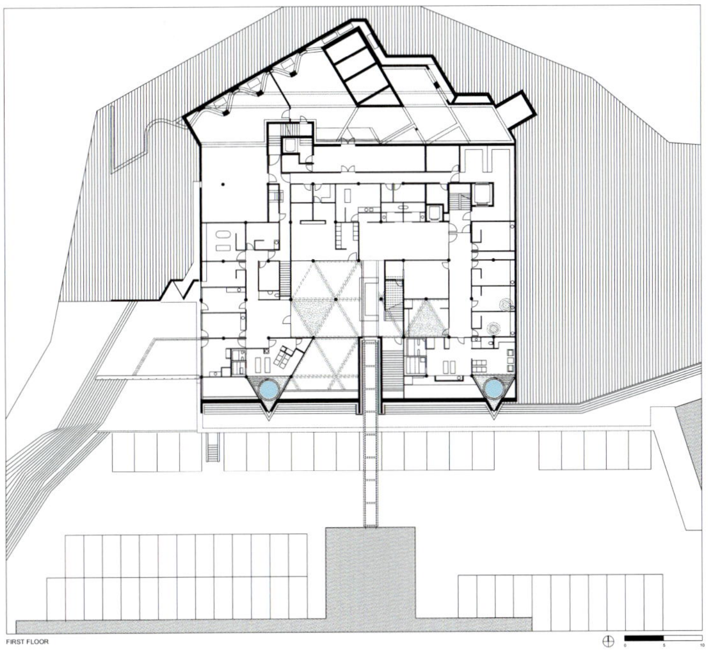

FIRST FLOOR

2

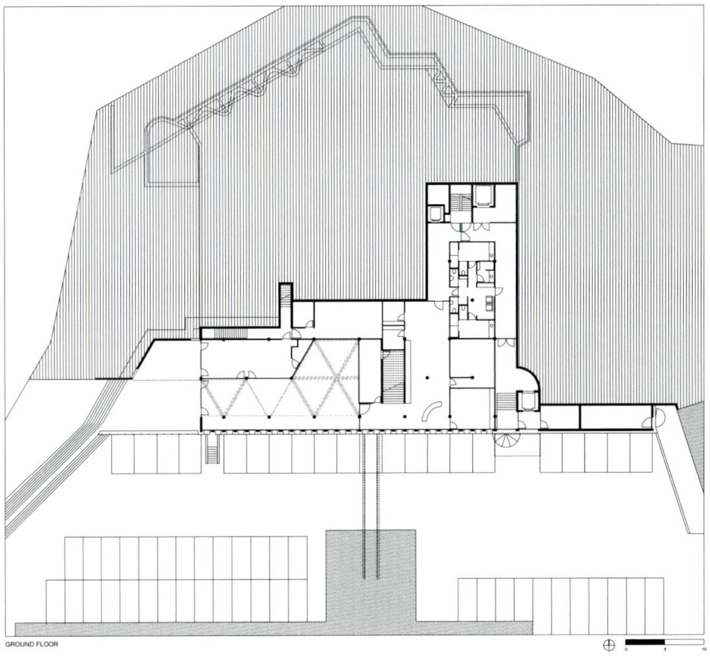

GROUND FLOOR

3

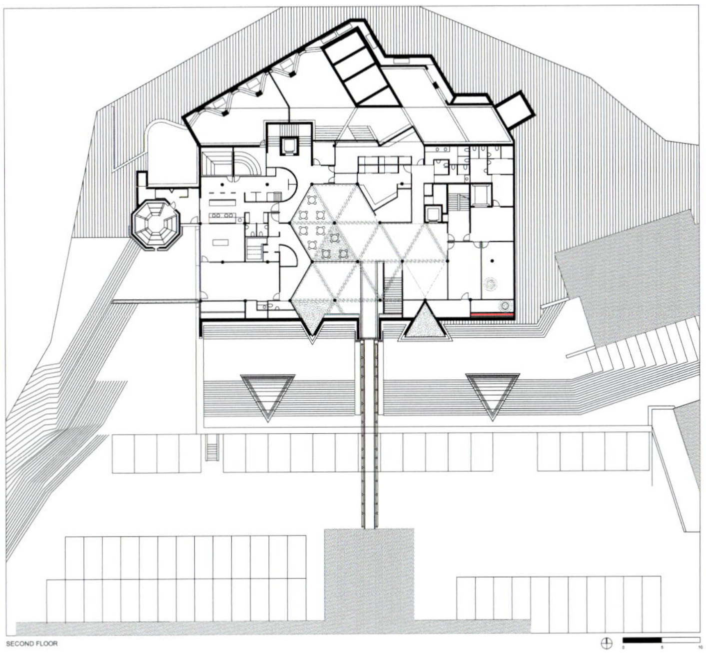

SECOND FLOOR

4

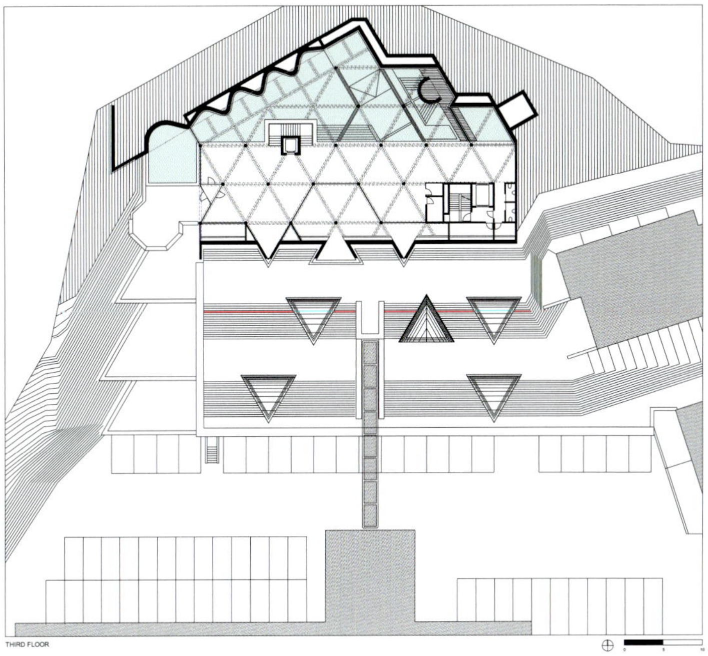

THIRD FLOOR

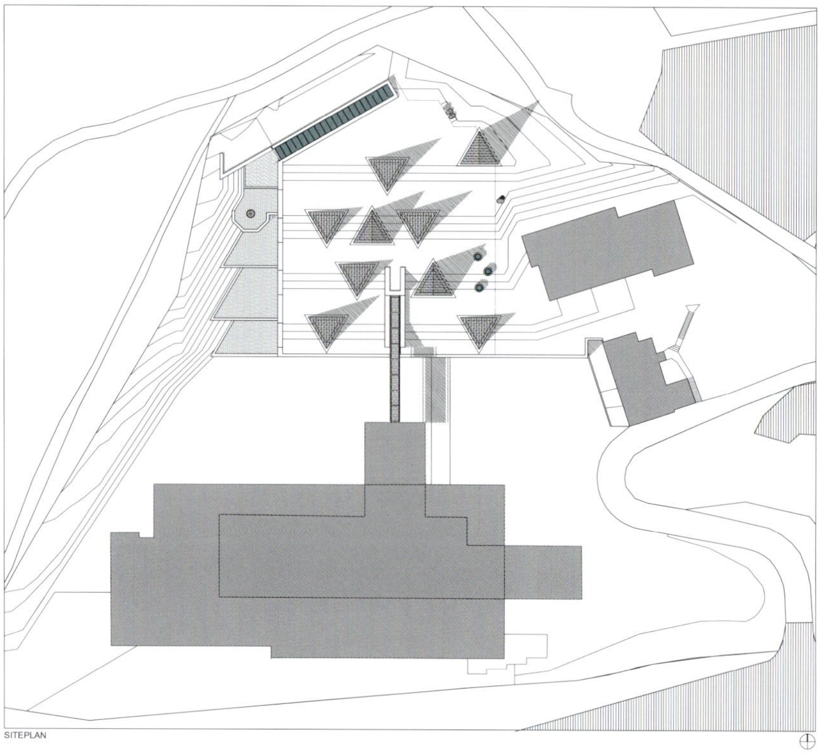

SITEPLAN

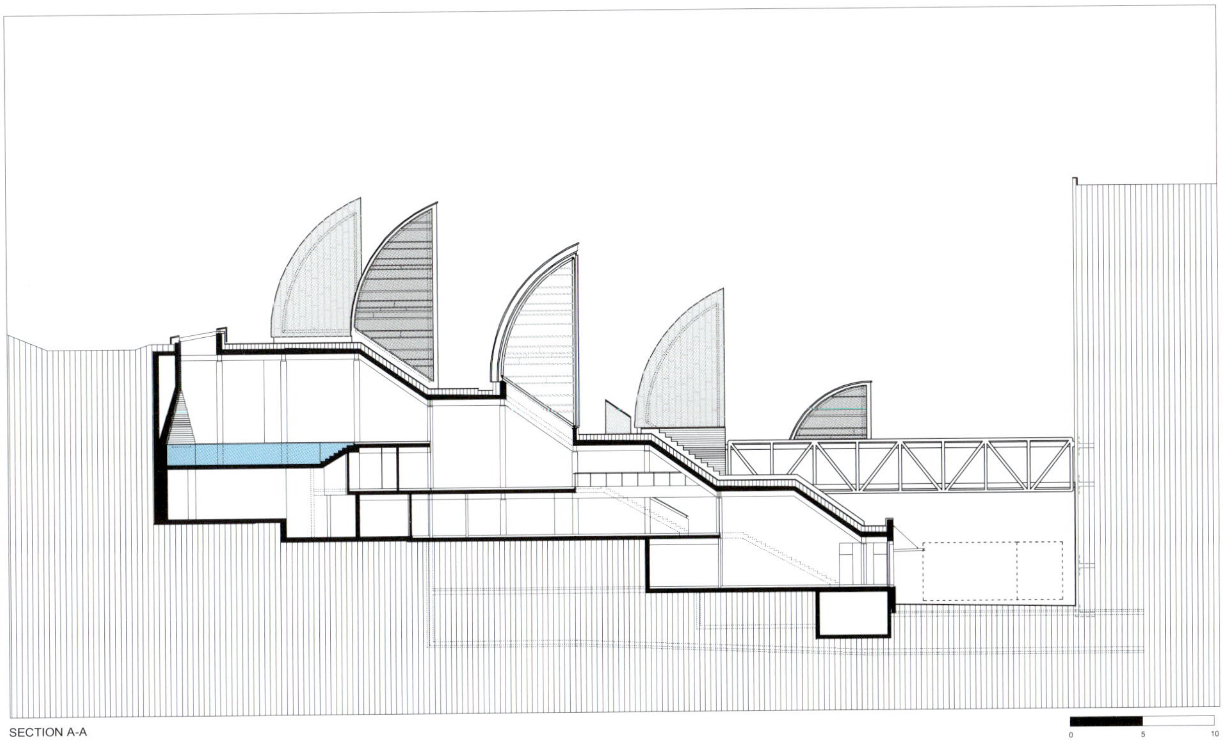

SECTION A-A

1.40 m

Spa 'Casa Velha do Palheiro'

Title: Spa 'Casa Velha do Palheiro'
Design Agency: MSB Architects
Location: Funchal, Madeira Island
Client: Hotel 'Casa Velha do Palheiro'
Area:950 m^2
Photography: Miguel Apolinario

MARSH & ASSOCIATES, INC.

Bella Collina

Title: Bella Collina
Design Agency: Marsh & Associates, Inc.
Location: Montverde, Florida, USA
Client: Ginn Clubs & Resorts
Area:6040 m^2
Photography: C. J. Walker

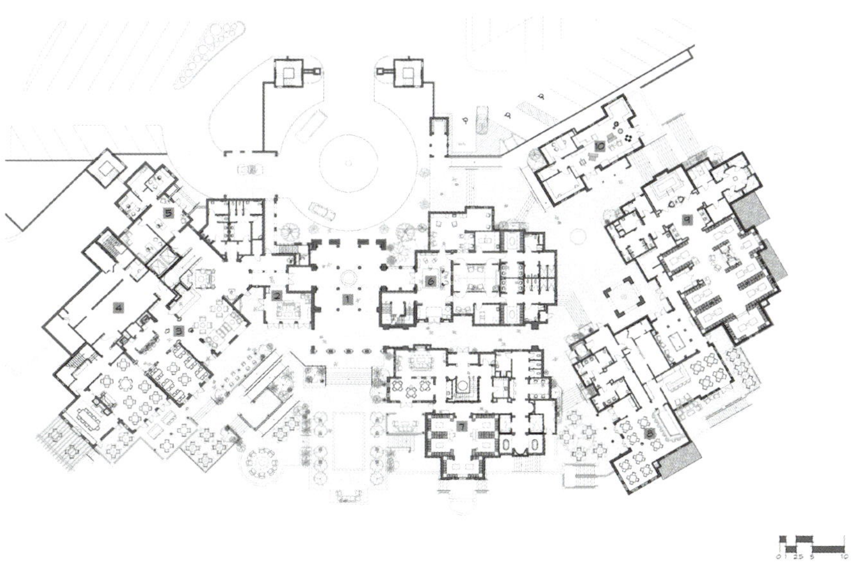

1

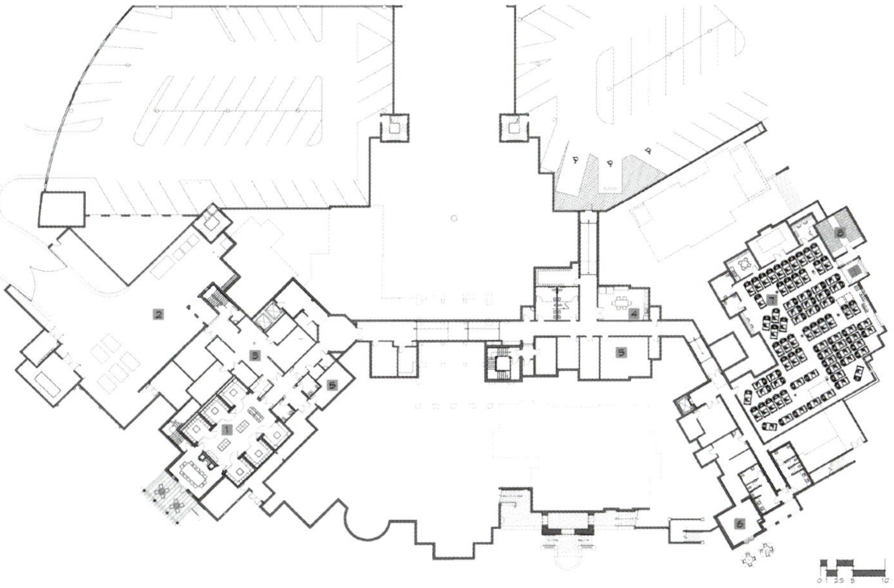

2

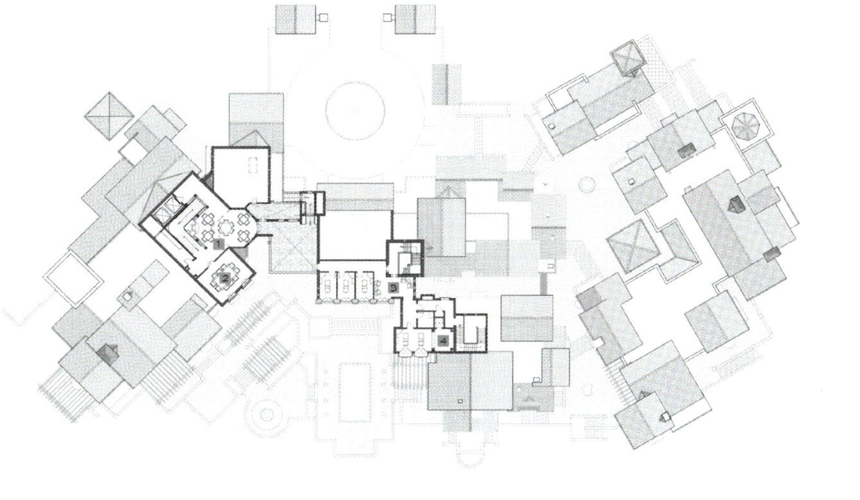

3

introduction:

The clubhouse at Bella Collina, a Tuscan-themed club and resort near Orlando, Florida, provides a spectacular blend of old-world style and abundant services. Its plan offers a variety of spaces and experiences: social and secluded, lively and reserved, opulent and casually elegant, and delivers them in delightful indoor and outdoor locations that rejoice in the surrounding golf course, lakes, vineyards and citrus groves. The layout and massing create the effect of a longstanding village built and refined over time. Inside this sense of history and authenticity continues, with vaulted brick ceilings, timber beams, cloisters, rich fabrics and colors, and a cavernous wine cellar. The result is an experience of comfort and opulence that appeals to the most discerning guest.

The clubhouse includes a full spa and salon that offer relaxing and gratifying treatments in warm and serene settings. The spa opens from the central courtyard, and provides a surprisingly contemporary architectural experience within an otherwise old-world setting. The main level includes lavish locker, steam room, and hydrotherapy facilities, around an enclosed relaxation area that is open to the sky above. The salon is here, as well, with opportunities for hair and skin treatments, manicures, and pedicures with a view! A chandelier-lit grand stair leads to the single and couples massage treatment rooms on the upper level, with solid copper soaking tubs and glass-paneled patio doors providing sweeping vistas across the lake. Finally, an enclosed 'bridge' leads female guests over the busy streetscape below to the luxurious Women's Lounge, allowing them discreet access to and from their spa appointments.

Golf Inc. magazine called Bella Collina's facility 'magnificent' in naming it their 'Private Clubhouse of the Year' for 2008.

The Club at Black Rock

Title: The Club at Black Rock
Design Agency: Marsh & Associates, Inc.
Location: Coeur d' Alene, Idaho, USA
Client: Black Rock Development, Inc.
Area:2880 m^2
Photography: Vickie Atkins

CYBEX

MAI
MARSH & ASSOCIATES, INC.

Cordillera

Title: Cordillera Ranch
Design Agency: Marsh & Associates, Inc.
Location: Boerne, Texas, USA
Client: Cordillera Ranch Development Corp.
Area:4000 m^2
Photography: Carlton Wade, Mark Knight

introduction:

Texas 'Hill Country', a region near the geographical center of this vast state, has a distinct architectural style that originated with the German immigrants and settlers of the mid-nineteenth century. The challenge for the design of the clubhouse, spa, and fitness center was to develop an authentic rendition of this distinctive style, in a strategic format that would enable phased development. The well-received solution offers a fresh but loyal interpretation of the rock-and-frame Hill Country vernacular. Walls and foundations are built of pale local limestone, found in abundance just beneath the region's thin topsoil. Long and shallow-pitched saltbox roofs of galvanized metal are supported on locally-harvested peeled cedar logs, whose durability is suited to the fickle climes of the area. Log frame and half-timbered upper levels and breezeway elements complete the hand-hewn historic feel.

The club's layout is that of a traditional stepped ranch, and its interior is one of rural comfort and ease. The squared plan is reminiscent of a German farmhouse, with multiple structures joined around a common courtyard, built sequentially. The result is a community of structures that embraces the features and folds of the site, in an arrangement that logically positions its functions around a lively shared core.

While this approach might suggest a lack of urgency uncommon in developments of the last decade, it is precisely the message of this club. The atmosphere is relaxed, the lifestyle unhurried. Membership continues to grow, with golf, fitness, therapy, and social activities flourishing. And in their simple comforts these rustic buildings convey the gentleman rancher's commitment to hard work and responsibility, community and heritage, and deference to the nurturing power of the land.

1

LEGEND
1 DINING
2 LOUNGE
3 KITCHEN
4 STAFF BREAK ROOM
5 LIVING ROOM
6 ADMINISTRATION
7 GOLF SHOP
8 WOMEN'S LOCKERS
9 MEN'S LOCKERS
10 SPA/TREATMENT
11 POOL EQUIPMENT
12 SNACK BAR
13 PAVILION
14 POOL
0 1 2.5 5 10

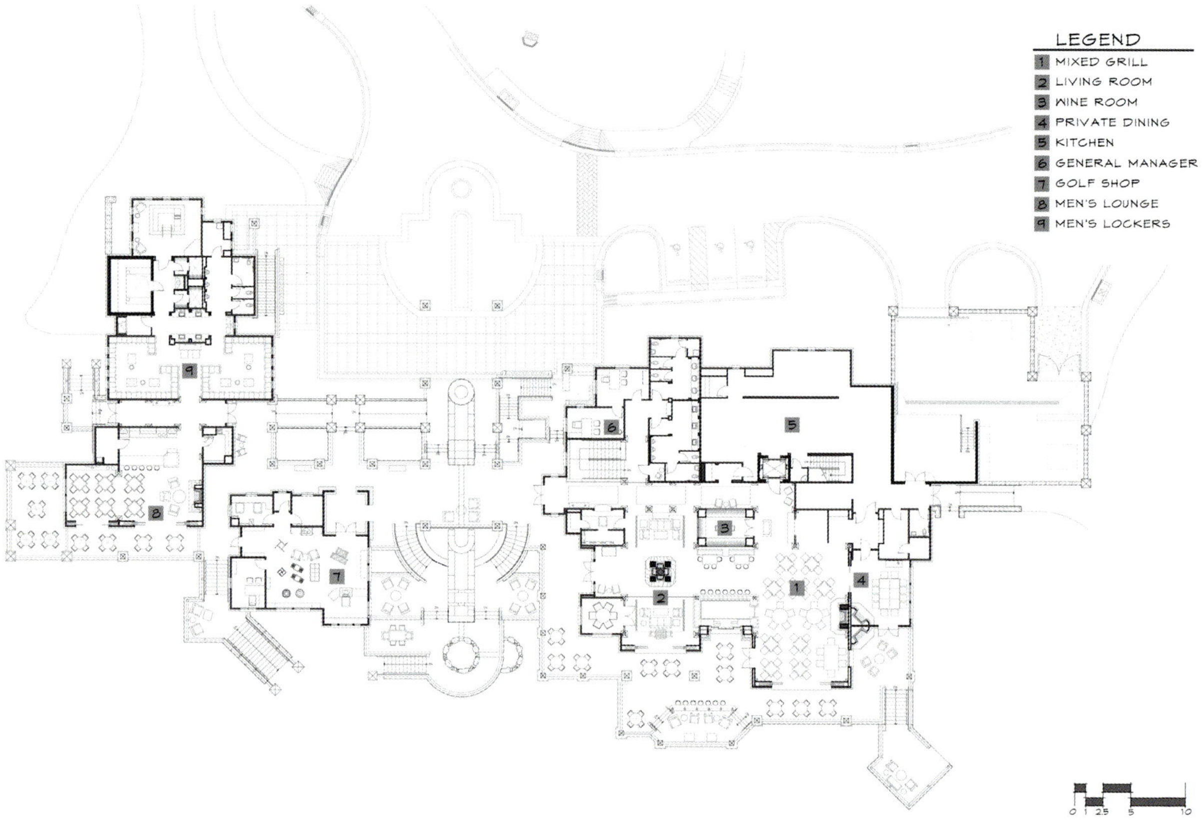
LEGEND
1 MIXED GRILL
2 LIVING ROOM
3 WINE ROOM
4 PRIVATE DINING
5 KITCHEN
6 GENERAL MANAGER
7 GOLF SHOP
8 MEN'S LOUNGE
9 MEN'S LOCKERS
0 1 2.5 5 10

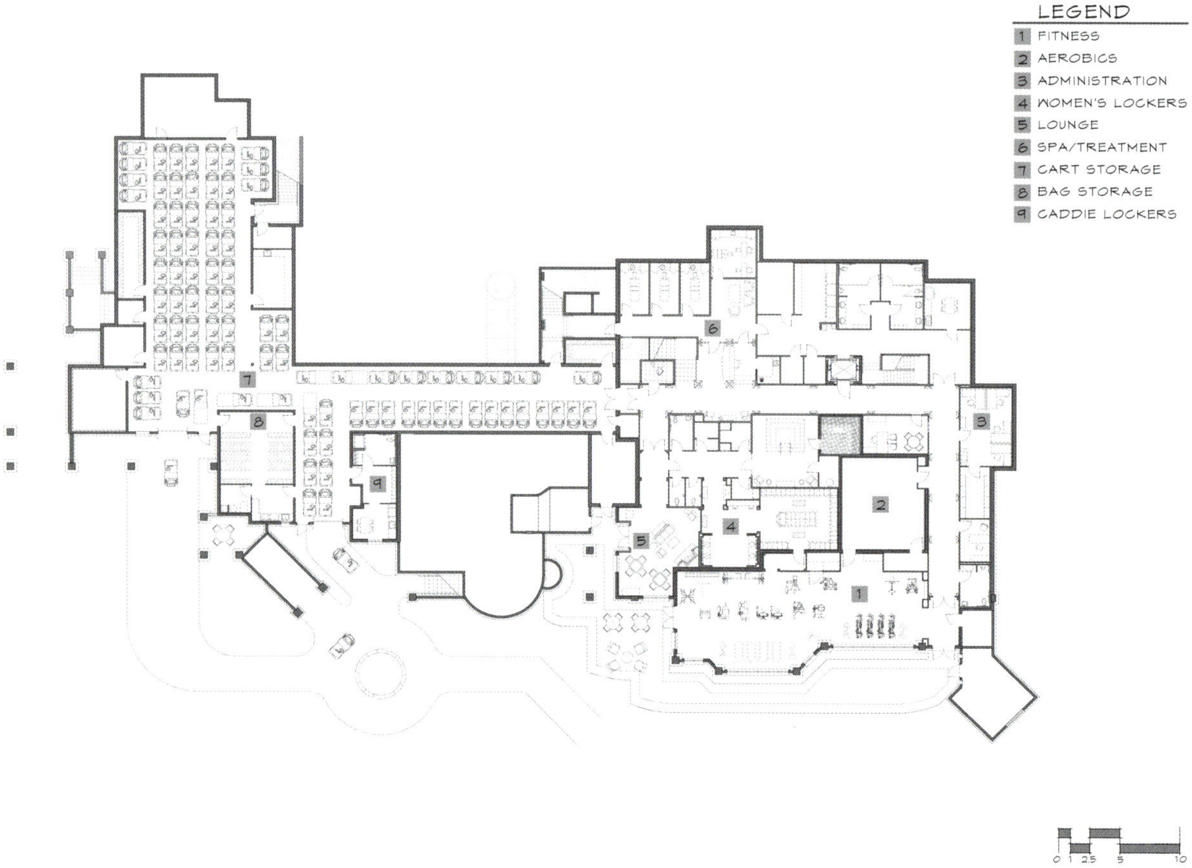
LEGEND
1 FITNESS
2 AEROBICS
3 ADMINISTRATION
4 WOMEN'S LOCKERS
5 LOUNGE
6 SPA/TREATMENT
7 CART STORAGE
8 BAG STORAGE
9 CADDIE LOCKERS
0 1 2.5 5 10

PINE CANYON
PINE CANYON
EXIT

SPINNER
SPINNER
SPINNER
SPINNER
SPINNER

Kirana Spa

Title: Kirana Spa
Design Agency: Bensley Design Studio
Location: Ubud, BALI, Indonesia
Client: SHISEIDO & PT Puri Saren Agung
Area:18,000m^2
Photography: Krishna Adithya

Introduction:

Located on the outskirt of Ubud, Kirana Spa is unparalleled spa facility. Winner of Spa Asia's 'Best Day Spa', Kirana Spa is a joint project between Shiseido and PT Puri Saren Agung, a hotel management company belonging to Ubud Sukawati royal family, which began receiving its first clients in early 2002. Since Shiseido realized that they needed a new image for this project, Lek Bunnag who was given responsibility to design the architecture and interiors and Bensley Design Studio gave verdict that Shiseido needed a new unhurried, natural, even mystical image for this project.

Spread over an area of 194,000 square feet, Kirana Spa was divided into two distinct garden areas : the upper gardens and the lower Ayung Gardens. The upper gardens are for the spa suites with all its facilities, than the remaining treatment rooms are in the lower gardens. The origin of the word 'Kirana' is 'Light' and 'Aura' in Sanskrit. It means 'Beauty beyond merely the physical' in Indonesian.

This spa has been designed to blend in seamlessly with the natural surroundings by use of tropical flowers and trees as an integral part of the spa, local materials, are concerned the Ayung Rivers as strength of the site to make a pleasant ambience by its water sounds.

In this spa, some concepts of the Japanese garden were adapted with Balinese old story of 'tirta hantu' for design of water flows over the gardens. A system that pumped the water up to the top of the gardens then it would make the effect of running water to be ever present, disappearing, and reappearing along the journey to the bottom of the valley by gravity in myriad ways. This system used many components of these gardens like walls, channel, waterspout, artworks to make many scenarios which have to be explored in person to understand and appreciate them.

Eco Spa

Title: Eco Spa
Design Agency: Spaplan
Location: Milan, Italy
Area:70 m^2

Introduction:

The Spa Mais Vida project was actually a remodeling of an existing edification, which was a Gym Academy previously. The main objective of this project was to reduce as much as possible the 'industrial' look inside of the building. There weren't any divisions among the spaces and these were opened, showing off the internal structure. By lowering the ceilings in different levels and by using different materials, we were able to modify the structure appearance and achieve a look pleasant and according to the client's needs.

The SPA is comprised of two floors: the first floor and the mezzanine floor. The property's total area is of 960m², being the first floor an excess of 600m² and the mezzanine floor of 360m². The SPA's reception area is a total of 110m², with a 5.1 m high headroom.

The mezzanine floor invites you to relax in a cozy lounge, looking out to the hills of Vale do Sereno, which area is of 47m² and headroom ranging from 2.5m to 3.6m in height. Inside the massage and aesthetics chambers, the headroom is 2.6m high, providing for more comfort to our guests. The high points of the building and surroundings have been enhanced, taking advantage of the first floor hall's headroom and the beautiful view of the Vale do Sereno, benefiting from neat shapes, superposed plans and neutral tones.

It is interesting to emphasize that the entire architectural project has been conceived, taking into consideration the desires of our guests. On top of that are personal techniques and tastes, deriving from our own Spa experience. With that in mind, researches have been carried out, leading to the aesthetic and functional consensus for the Spa. Contemporary Asian, European and also Brazilian inspirations, allied to the continuous responsibility for the environment, are milestones of the project, securing its austere and stout aspect, despite using only simple and neat components.

The two month time-frame of the construction was also decisive as the project was being elaborated. The chronogram had to be very well calculated and defined before hand, lest mischances occurred, once the Spa's renovation work started.

Since this was a project to be completed in a short term, we aimed at minimizing most significant interventions, yet capturing changes that would transform the warehouse and gym, formerly establish therein. Lowered plaster linings required massive metal structures as supports, regardless of the existing lining, since those would not withstand overloads.

A few walls received upscale brickwork, benefiting from the use of drywall technique, which permitted a significant gain in work, besides creating fragmented environments and varied headroom setting, with indirect lighting.

At the relaxing lounge, the use of mild tones, mossô-bamboo seedlings, water mirrors and other mirrors provided for an optimized view of the hills, for more tranquility and relaxation. The staggered lining (due to the bowed shape of the warehouse) and the indirect and diffused lighting provide for a welcoming sensation.

In the entrance hall, Interfloor PVC stripe-coated gates are also welcoming and evoke comfort and stateliness. The hall is illuminated by exquisite Scatto overhangs, conducting to a lovely walk across the corridor that looks out to the Vale do Sereno Hills. There, guests will also be delighted to pay a visit to the Studio Power Plate and the Lounge Café. Both are privileged by a glass wall, which guarantees a natural ventilated and illuminated environment.

The façade, nearly entirely covered with cement boards, gained added quality and was built in a timely fashion, besides generating little civil construction residual debris. Boards and bricks were finished in a coppery paint, reminiscent of the ore mountains of the State of Minas Gerais.

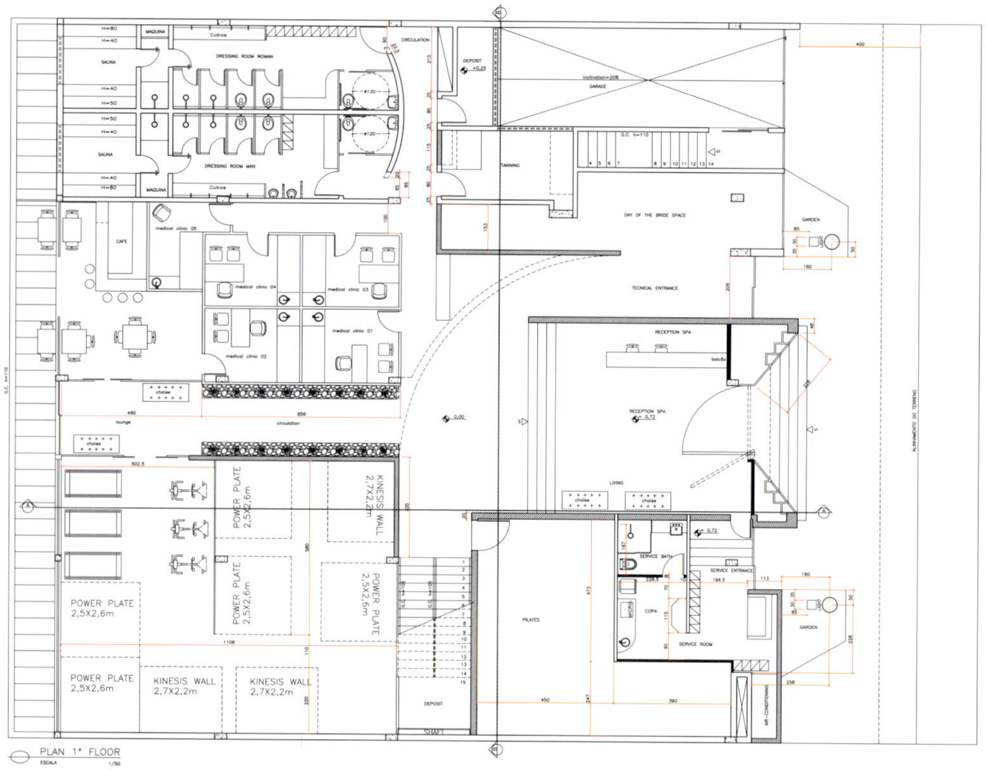

PLAN 1° FLOOR
ESCALA 1/50

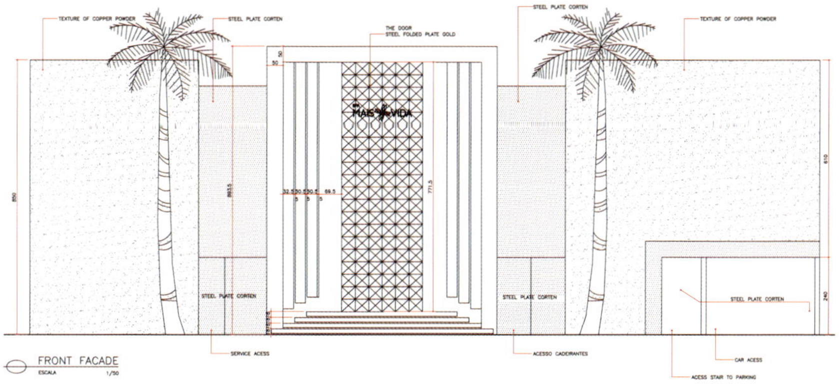

FRONT FACADE
ESCALA 1/50

SECTION AA
ESCALA 1/50

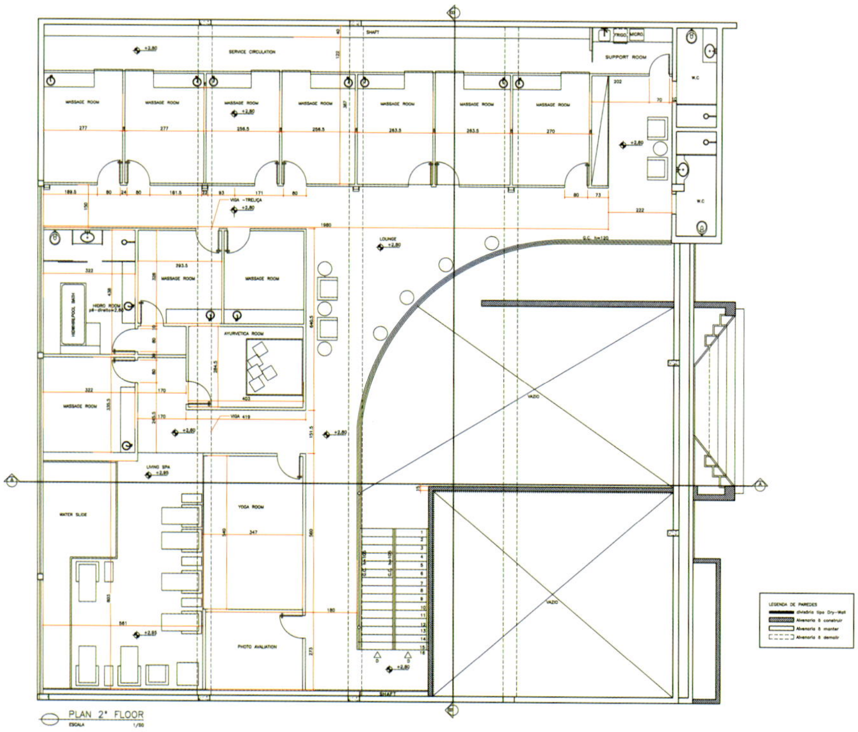
SERVICE CIRCULATION
SHAFT
SUPPORT ROOM
MASSAGE ROOM
W.C
VIGA - TRELIÇA
LOUNGE
HIDRO ROOM
AYURVETICA ROOM
VIGA
VAZIO
LIVING SPA
WATER SLIDE
YOGA ROOM
PHOTO AVALIATION
LEGENDA DE PAREDES
divisória tipo Dry-Wall
Alvenaria à construir
Alvenaria à manter
Alvenaria à demolir
PLAN 2° FLOOR
ESCALA 1/50

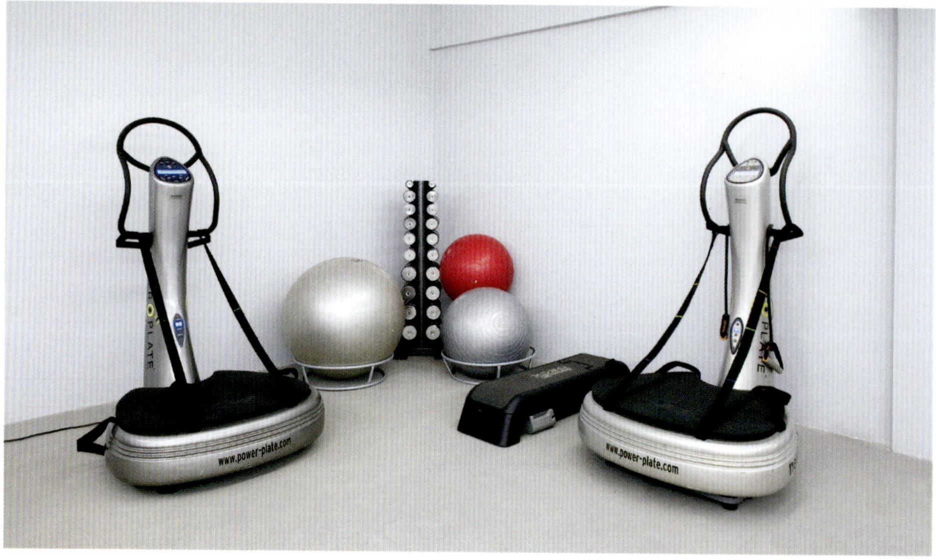
www.power-plate.com
www.power-plate.com

Xi Er Si Chi Dian

Title: Xi Er Si Chi Dian
Design Agency: Da Zhan Design Consultant Limited.
Location: Shenyang, China
Area: 18,000 m²
Photography: HK Ata Cinema Industry Limited

introduction:

Xi Et Si is the transliteration of 'Health'. It is aimed at providing health and leisure. The whole project blends in exotic characteristics, South-Asia elements and the palace style from the Middle East. Through arranging and blending materials, models, and symbols, the designer fully interprets the theme of 'Luxurious Leisure'.

5

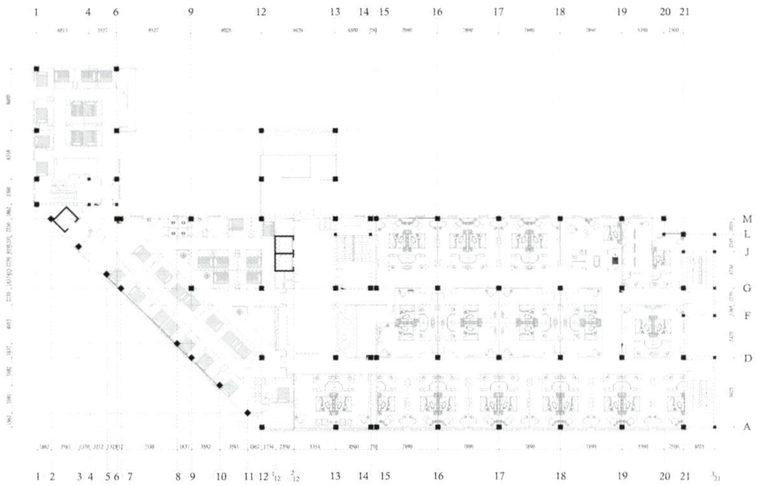
1 4 6 9 12 13 14 15 16 17 18 19 20 21
S R Q P M K H G E D C B A
M L J G F D A
1 2 3 4 5 6 7 8 9 10 11 12 13 14 15 16 17 18 19 20 21

6

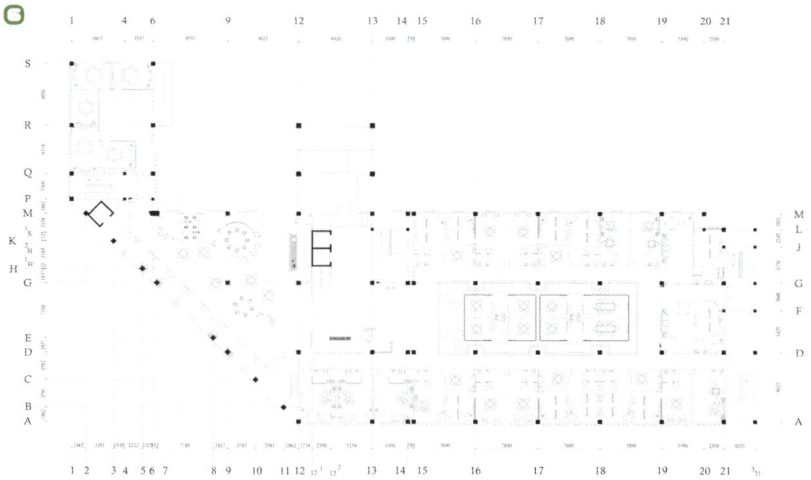
1 4 6 9 12 13 14 15 16 17 18 19 20 21
S R Q P M K H G E D C B A
M L J G F D A
1 2 3 4 5 6 7 8 9 10 11 12 13 14 15 16 17 18 19 20 21

7

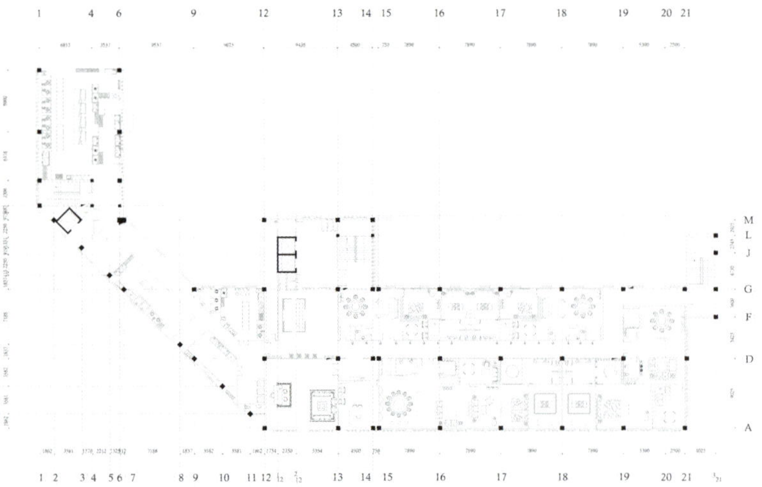
1 4 6 9 12 13 14 15 16 17 18 19 20 21
S R Q P M K H G E D C B A
M L J G F D A
1 2 3 4 5 6 7 8 9 10 11 12 13 14 15 16 17 18 19 20 21

Angsana Spa Tivoli Marina Vilamoura

Title: Angsana Spa Tivoli Marina Vilamoura
Design Agency: Atelier de Arquitectura (ADA)
Location: Portugal
Client: Angsana Spa
Area: 880 m^2

introduction:

Angsana Spa greets spa enthusiasts in mainland Portugal for the first time in Tivoli Marina Vilamoura, a reputed hotel for luxury tourism in the Algarve. Overlooking the spectacular Atlantic Ocean, Tivoli Marina Vilamoura is located next to Portugal's finest marina and attracts both holidaymakers and business delegates due to its cosmopolitan charm. The interior design presents the perfect combination of modern Asian elements and European Algarve style. Angsana Spa unveils 11 treatment rooms designed with contemporary splendor and serenity to complement the authentic spa experience. Home to a splendid range of sensory treats, Angsana Spa offers a range of award-winning treatments to spa aficionados.

ANGSANA
Spa

introduction:

The green tree of spa brand located in the center of entrance of reception guest hall and the roof is hung with pure crystal leaf, where you have a rest and change your mood. And there is care products exhibition to offer guests with spa products back home to use and extend brand promotion in the reception center. Passing the corridor that is made from constantly four-seasons changing LED is the highlight of lighting design. With the gorgeous light in the roof to come to VIP room, each room is equipped with independent dress up area, rest area, steam room and toilet and bath so as to reach a new generation of spa high demand allocation goal. Hence, designers apply water and soft light to the bath and bring the nature into interior space.

Sculpture leaves glass is separated from massage area and changing area. Wood patterns are painted on white pure wall. Moreover, the space is covered with the breath of nature and modern in the unceasingly visual and lighting. Yet the floor is used man-made stone low carbon environmental protection, also extended protection environment theme. The designer hopes to use the curve of the human body as inspiration. Therefore, adopting arc shape in the space and line to make guest have a sense of safety and identity and completely relaxed.

The spa consists of three sectors, including the swimming pool, the dressing room and gymnasium. The aerial swimming pool with constant temperature is unique for that shinning daylight. It would feel like be roaming in the heaven for its transparency and boundless of the pool wall while swimming in it. Besides, the overflowing water completely combines the interior waterscape of the hotel. 'Ripple' would be the design motif of the gymnasium. The woodiness modeling of top surface and wall just like the water rings as to emerge some warmth and rhythm. What's more, the 26-meter-long gym center in the air applies an infinite and outspread space by using some refractive makings in the background. By viewing the bustling cityscape, this area grants some authentic feelings to people those taking outdoor exercises.

WELCOME

Dolder Grand Spa

Title: Dolder Grand Spa
Spa Design: SPAd (Sylvia Planning And design)
Architect: Lord Norman Forster
Interior Design: United Designers London
Location: Zurich,Switzerland
Area:4000 m^2

introduction:

Pondering the grandeur of the Swiss alpine peaks, passes, and valleys ignited Sylvia Sepielli's creative spirit in the development of the Dolder Grand Spa. The architecture, inspired by the beauty of the Swiss Alps, is punctuated with contemporary and Japanese influences that drive both facility features and the overall guest experience. At the Dolder Grand Spa, great attention has been given to the fusion of nature & modern advances, the past & present, and the blending of cultural healing practices, resulting in a balance that promotes wellness.

The Dolder Grand Spa is designed as a canyon, with a 'core' at the center of the building that all spaces radiate out from. The guest begins with an elevator ride to the spa, this journey represents the decent into a canyon. The core of the building signifies rejuvenation and personal discovery, and it is where the meditation area is located.

Specialty features that represent the Japanese influences help guests to prepare for their treatments. The Kotatsu is a footbath area in the wet area where guests can relax while having a cup of tea. The Sunaburo are Japanese-inspired warm pebble baths that guests lay in while enjoying the expansive vista overlooking Zurich.

Favorite areas to enjoy at the Dolder Grand Spa are the meditation area, library, ladies' & men's spa, fitness, aqua zone, chill out, café, and treatment spaces.

The Dolder Grand

Shop
Hairstyling
Spa Library
Toilets
Work out
Movement Studio
Mind-Body Studio
Outdoors

Shenzhen Jingyuan Yoga Studio

Title: Shenzhen Jingyuan Yoga Studio
Design Agency: Lonson Design Co.,Ltd
Designer: Yuan Jing
Location: Shenzhen, China
Area: 450 m^2
Main Materials: Black Galaxy marble, hatchway, mirror finished stainless steel, dope, etc.

Introduction:

This case is for Shenzhen Jingyuan Yoga Studio, located in Shenzhen Haiancheng mansion, with an area of about 450 sqm. It has function areas like the reception desk, products exhibit areas, big and small training rooms, meditation rooms, resting gardens, etc. And to make more space for training rooms, the antechamber, discussion areas and products exhibit rooms are combined as a whole in brief forms and colors of black and white, which makes the space generous and clean. And there is also added another kind of feeling with selections of Chinese classical furniture.

The two training rooms meet the needs of different trainees. With ceilings and lights designed with the intrinsic features of original girders, light can be changed accordingly making sure that trainees have a comfortable quiet atmosphere. In the smaller room, there is a mirror, which not only helps trainees see their movements but extends the space. While the two rooms are in the same kind of style and color, there are still differences, especially with their distinct decorations and furnishings.

Over the two square training rooms, there is an elliptical meditation room making use from wooden slat, which makes the room gentle and fragrant. It is like a finishing touch enhancing interestedness.

The big outdoor garden, which is originally a balcony, is also a hardly-seen beautiful scenery. Trainees there can lie fallow or have a taste or chat with others with a view of beauteous city and sea.

Shenzhen In Yoga GYM

Title: Shenzhen In Yoga GYM
Design Agency: Shenzhen Horizon Space Design Co.,Ltd
Designer: Han Song
Location: Shenzhen, China
Area: 1000 m^2
Photography: Han Song

introduction:

This case is located in Shenzhen Futian district COCO park's 3F. Stylist Han Song dig up the inner essence of design as always. Actually anything's external forms are representations, what's important is what its inner essence can bring you. This is the consensus with owner at the beginning of design in yoga clubs,it's not only the source of my yoga museum design,but also her way of business. In other words,what's urbanite need: Power? Money? Fulfillment? Coming here just for fitness or consumption? I think these are not exactly. In the increasingly busy, pale, modern urban life, they may need more a kind of humane care, a kind of remote leisurely happiness.

Stylist Han Song assembled metope himself with abandoned old woodcarving and rotten wood collected from timber mill at the entrance, the cultural precipitation of the metope is the only, other contemporary adornment gimmick can't express it. At the same time this relaxed expression reminds a cheerful feeling of urbanite long memory who living in the immigration city.

Rest area directly connects with outdoor garden. You can brush away the yarn and walk barefoot from indoor to the soft outdoor floor. This is a fantastic feeling, it can relax you completely to listen to the murmuring sound of water. The whole wall of bamboo in the Yoga room creates an atmosphere of oneness in a poetic way.

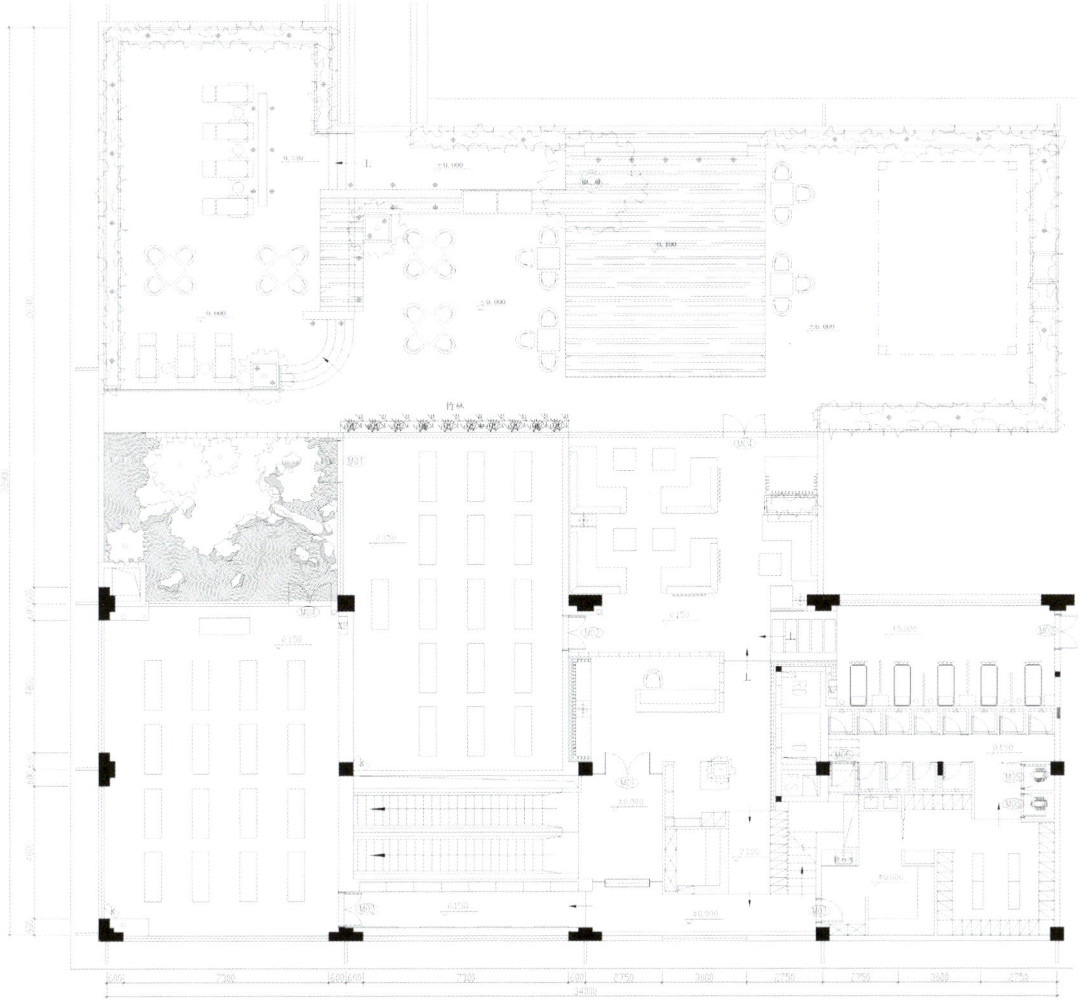

1 : 100

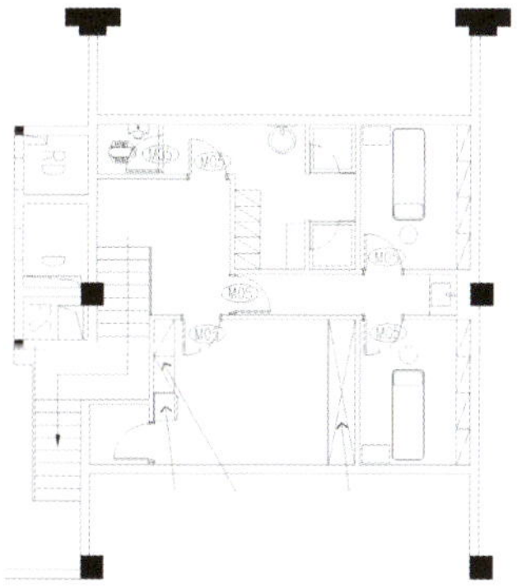

1 : 100

Acknowledgement

We would like to thank all the designers and companies who made significant contributions to the compilation of this book. Without them, this project would not have been possible. We would also like to thank many others whose names did not appear on the credits, but made specific input and support for the project from beginning to end.

Future Editions

If you would like to contribute to the next edition of Artpower, please email us your details to: info@artpower.com.cn.